JOINT SPECIAL OPERATIONS
JSOU
UNIVERSITY

SOF Role in Combating Transnational Organized Crime

Essays By

Brigadier General (retired) Hector Pagan
Dr. Emily Spencer
Mr. Mark Hanna
Brigadier General Mike Rouleau
Colonel Earl Vandahl
Dr. Peter McCabe
Professor Celina Realuyo
Colonel Bernd Horn
Dr. Christian Leuprecht
Colonel Bill Mandrick, Ph.D.
Mr. Randy Paul Pearson
Mr. Michael Miklaucic

Edited by

Colonel (retired) William Mendel
Dr. Peter McCabe

The JSOU Press
MacDill Air Force Base, Florida
2016

SOF Role in Combating Transnational Organized Crime

Published in the United States by Joint Special Operations University Press.

Press publications are available for download from the JSOU Library web page located at https://jsou.libguides.com/jsoupublications. Comments about this book are invited and should be forwarded to the Director of the Center for Special Operations Studies and Research, Joint Special Operations University, 7701 Tampa Point Blvd., MacDill AFB, FL 33621.

****

The JSOU Center for Special Operations Studies and Research (CSOSR) is currently accepting written works relevant to special operations for potential publication. For more information, please contact the CSOSR Director at jsou_research@socom.mil. Thank you for your interest in the JSOU Press.

Sign up to receive notification of new JSOU Press publications with a GovDelivery subscription, available at: https://public.govdelivery.com/accounts/USJSOU/subscriber/new.

****

ISBN 978-1-941715-05-5

Book cover design by Ms. Lana Burroughs, JSOU

Printed in the United States of America

Recent Publications of the JSOU Press

The Collapse of Iraq and Syria: The End of the Colonial Construct in the Greater Levant, December 2015, Roby C. Barrett

Tactical Operations for Strategic Effect: The Challenge of Currency Conversion, November 2015, Colin S. Gray

Countering Violent Extremism in Mali, November 2015, Mark Moyar

The 2005 Iraqi Sunni Awakening: The Role of the Desert Protectors Program, October 2015, William Knarr

2016 Special Operations Research Topics

Saudi Arabia: Modernity, Stability, and the Twenty-First Century Monarchy, June 2015, Roby C. Barrett

Special Operations Forces Reference Manual, Fourth Edition, 2015

2015 Special Operations Essays, May 2015

Building Partner Capacity, February 2015, Harry R. Yarger

Islam: Ideology and Conflict, December 2014, Roby C. Barrett

Village Stability Operations and the Afghan Local Police, October 2014, Mark Moyar

Contents

Foreword

Introduction

The Strategic Environment

Policy and Strategy

Operations

Foreword

How do SOF Fit into Fighting Transnational Organized Crime

Rear Admiral Kerry Metz

It is my pleasure to present this publication's summary and conclusions from our 2015 Symposium on Transnational Organized Crime (TOC) conducted with our Canadian Special Operations Forces (CANSOF) and Mexican special operations and law enforcement partners. Why is TOC a concern and why should it be a concern to our special operations community? The nexus between criminal and terrorist networks is significant and evolving, and the threat to our nations' security demands that we collectively explore regional whole-of-government approaches and determine the potential Special Operations Forces (SOF) roles for countering and diminishing these violent destabilizing networks. The panel summaries that fill this publication detail our national sovereignty challenges, shared opportunities, and most importantly, some very useful conclusions regarding the employment of SOF against TOC.

For this years' symposium, we are again indebted to U.S. Special Operations Command's Joint Special Operations University for establishing an academic environment with subject matter experts so we could collaboratively explore regional implications. CANSOF Command's recent publication, '*By, With, Through*' captures the theme and discussions from our 2014 symposium and highlights the imperative of working with national and agency partners to solve complex transnational problems. Because of the groundwork we established in 2014, we were able to focus this year's symposium not only on the specific Western Hemisphere TOC challenges, but also on potential roles and unified actions for SOF. The wonderful diversity of profession, experience, and expertise of our panelists and attendees enabled focused discussions that were thought provoking and extremely useful. Our discussions focused on our shared national interests, our unique perspectives, and arguably our disjointed national policies. Most important to United States Special Operations Command North (SOCNORTH), and I hope for our partners, were our discussions on how the employment of SOF could address TOC challenges.

Let me conclude by expressing my appreciation, gratitude, and *merci* to our Canadian SOF partners who continue to make these symposiums relevant and impactful to SOCNORTH and to our greater SOF community. Equally, I want to extend my appreciation and *gracias* to our Mexican partners who attended our symposium. Without their insights and operational perspectives, this symposium would have faltered. And finally, I want to extend our most sincere gratitude and shared appreciation for our SOF and law enforcement professionals who engage without hesitation, often at great sacrifice, and with great skill to preserve our nations' sovereignty, way of life, and prosperity in our hemisphere.

Kerry M. Metz
Rear Admiral, U.S. Navy
Commander, Special Operations Command North

Introduction

The SOF Role in Combating Transnational Organized Crime

Colonel (retired) William (Bill) W. Mendel

In April 2015, military and civilian personnel from Canada, Mexico, and the United States came together at Colorado Springs, Colorado, for a symposium hosted by U.S. Special Operations Command-North and facilitated by Joint Special Operations University (JSOU) and Canadian Special Operations Forces Command (CANSOFCOM). Their task was to examine the role of Special Operations Forces (SOF) in combating transnational organized crime (TOC). After opening remarks by Rear Admiral Kerry Metz, a U.S. Navy SEAL and SOCNORTH commander, the panelists and plenary participants set to work considering a wide range of issues attending to the TOC threat. Symposium panelists and speakers synthesized the results of their research and panel discussions about TOC, and these are found in the chapters of this report of proceedings, *The SOF Role in Combating Transnational Organized Crime.*

So what is TOC? President Obama in his strategy to counter TOC advances this description:

> Transnational organized crime refers to those self-perpetuating associations of individuals who operate transnationally for the purpose of obtaining power, influence, monetary and/or commercial gains, wholly or in part by illegal means, while protecting their activities through a pattern of corruption and/or violence, or while

---

Bill Mendel is a retired U.S. Army colonel and senior fellow with the Joint Special Operations University. He serves on the Editorial Advisory Board of the JSOU Press, supports senior fellow research activities, and leads seminars and exercises. He has published numerous articles and studies concerning national security, strategy, and campaign planning.

> protecting their illegal activities through a transnational organizational structure and the exploitation of transnational commerce or communication mechanisms.[1]

The United Nations (UN) defines TOC as a:

> structured group of three or more persons, existing for a period of time and acting in concert with the aim of committing one or more serious crimes or offences … in order to obtain, directly or indirectly, a financial or other material benefit.[2]

Douglas Farah, author and veteran analyst of financial crime, armed groups, and TOC, focuses upon TOC as drug trafficking, money laundering, and human trafficking, plus trafficking in weapons of mass destruction.[3] Compiling a complete list of TOC activities is challenging.

The TOC Threat

The scope of the TOC threat is consuming: "fifty-two activities fall under the umbrella of transnational crime."[4] Drug trafficking tops the list and it brings in about $320 billion a year; human trafficking is worth about $32 billion; and gunrunning earns about $300 million annually.[5] Just moving all this illicit money around the globe is lucrative too. Money laundering involves 2 to 5 percent of global gross domestic product (GDP), or $800 billion to $2 trillion in current U.S. dollars on which banks make a commission.[6] More important than the money involved, the human cost is horrific, as is the impact on weak and struggling states that become corrupted and cannot provide basic governance and services for their countrymen.

Weak or failing states and attendant TOC are identified in the 2015 U.S. National Security Strategy (NSS) as among the top strategic risks to the country because of the significant security consequences. Weak states provide the sanctuary for the crime-terror-insurgency nexus to flourish. William F. Wechsler, former U.S. Deputy Assistant Secretary of Defense for Counternarcotics and Global Threats, provides these examples: In Afghanistan "the Taliban continues to receive a large percentage of its revenue from the heroin trade." And in Lebanon "Hizballah has become a drug trafficking and money laundering organization as well as a terrorist group."[7] This is a reminder of the early 1990s when the Revolutionary Armed Forces of Colombia (FARC) engaged in TOC (drug trafficking, gunrunning, kidnapping, and extortion) to fund its "big guerrilla business."[8] A military role for counter transnational organized crime (CTOC) strategy was made apparent in the execution of "Plan Colombia." Secretary Wechsler recalls:

> In Colombia, the Department's sustained counternarcotics and security assistance delivered military training, tactical and operational support, capacity building on intelligence sharing and information operations, equipment, and human rights training.[9]

Thus, TOC is a longstanding threat to the interests of democratic nations. TOC is deeply rooted in the preconditions for terrorism and insurgency and the thirst for power and wealth, as well as in the policies of nations that make it profitable.[10] Given TOC as a strategic threat, it is fitting for the international military community, especially the SOF community, to examine the role of special operations in supporting the CTOC effort. This is nothing new, as SOF elements have been involved in countering some dimensions of transnational crime since the early days of the drug war.

The Backstory

Military support to counter transnational crime can be traced to the U.S. Anti-Drug Abuse Act of 1988 (10 USC 1502), which established the creation of a drug-free America by 1995 as a U.S. policy goal.[11] The resulting National Drug Control Strategy outlined two major campaigns: supply reduction and demand reduction. To reduce the supply of drugs entering the United States, the major effort was to stop illicit drugs overseas and in transit.[12]

The Drug Enforcement Administration (DEA) and the military played the major roles in the overseas actions. Defense Secretary William J. Perry directed that the Department of Defense (DOD) would provide support for the detection and monitoring of the transport of illegal drugs, provide security assistance in source nations, and would support the DEA's Kingpin Strategy in "Dismantling the Cartels."[13] At the time of the first National Drug Control Strategy of 1989, Joint Task Force (JTF) 4, under Atlantic Command located at Key West and JTF-5 (Pacific Command, Coast Guard Island, Alameda, California) were established to fulfill the military's obligation to conduct air and sea detection and monitoring.[14]

At this same time, DOD established JTF-6 (now JTF-North under U.S. Northern Command) at El Paso, Texas. Billeted in an old military jail, JTF-6 provided DOD support to drug law enforcement agencies in the Southwest border area. Typical missions of JTF-6 were to provide intelligence analysis, ground radar sensing, airborne reconnaissance, ground and air transportation, engineer operations, military exercises, ground reconnaissance, and mobile training teams.[15] Guided by a SOF liaison office on staff, the SOF role was to provide training assistance and ground reconnaissance. JTF-North continues this mission today with the new guidance to "support our nation's federal law enforcement agencies in the identification and interdiction of suspected transnational criminal organization (TCO) activities conducted within and along the approaches to the continental United States."[16]

It is well established in this volume that there is a military role in countering TOC, and within that role there is a place for SOF support. But with the advent of the new NSS and a supporting national Strategy to Combat Transnational Organized Crime (SCTOC), what roles have these new strategic policy documents carved out for SOF?

The Strategic Guidance

The NSS identifies TOC as a "top strategic risk to our interests," asserting there are "[s]ignificant security consequences associated with weak or failing states (including … transnational organized crime)."[17] One strategic objective, "build capacity to prevent conflict," relates to a potential SOF supporting role. The strategy explains that weak governance and grievances allow extremism to flourish and therefore:

> [t]he focus of our efforts will be on proven areas of need and impact, such as inclusive politics, enabling effective and equitable service delivery, reforming security and rule of law sectors, combating corruption and organized crime, and promoting economic opportunity …[18]

This suggests the possibility for employing a number of SOF core activities in the context of capacity building: preparation of the environment, direct action, security force assistance, military information support operations, civil affairs, and perhaps more. Examples of likely SOF core operations could include counterinsurgency, unconventional warfare, foreign internal defense (FID), and counterterrorism (CT).[19] The strategy makes no specific mention of employing military power, but the president intends to "draw on all elements of our national strength" to accomplish his agenda, and presumably this would include SOF power.[20]

The national SCTOC lists 56 strategic concepts (called priority actions) in the strategy, and two clearly suggest a military role. First, the U.S. will "strengthen interdiction efforts in the air and maritime domains."[21] Much of this would be under the synchronizing and integrating role of U.S. Southern Command's Joint Interagency Task Force South, a military-interagency interdiction headquarters in Key West.[22] This involves the operations of the U.S. Coast Guard, Navy (with law enforcement detachments aboard), U.S. Customs Air/Border Patrol (for terminal interception in the continental U.S.), and other agencies. The role of SOF here is conceivable, but likely limited to an extreme situation such as countering weapons of mass destruction in a takedown at sea by Navy SEALs. But there is nothing new here, and Joint Interagency Task Force South procedures are well established through 26 years of interdiction experience.

Secondly, the TOC strategy intends to "disrupt drug trafficking and its facilitation of other transnational threats." To do this it will "leverage assets to enhance foreign capabilities, including CT capacity building … [and] … military cooperation."[23] Here is where SOF can make a significant contribution by its experience with FID missions that support a host nation's internal defense and development program.

The SCTOC is silent about the kinds of stateside operations that could be conducted by SOF. Nothing is said of Joint Task Force North, which is:

> the U.S. Northern Command organization tasked to support our nation's federal law enforcement agencies in the identification and interdiction of suspected Transnational Criminal Organizations' (TCOs) activities conducted within and along the approaches to the continental United States.[24]

JTF North integrates and coordinates DOD support to federal, state, and local drug law enforcement agencies throughout the United States. In past years SOF have provided training assistance, reconnaissance, and rapid response capabilities to JTF North, and this is likely to continue under the SCTOC.

Future TOC Missions

Given the NSS and SCTOC policy guidance, it is a safe bet that the combating TOC support missions assigned to the military, including SOF, will devolve from the counterdrug paradigm of the past.[25] This will possibly include some CT and counterinsurgency tasks in the admixture as we saw in Colombia.

The SCTOC is intended to complement and interlock with the National Drug Control Strategy, but more than that, the U.S. drug war and TOC war have been conjoined. The Office of National Drug Control Policy (ONDCP) will co-chair (with the NSC staff) the Interagency Policy Committee on Illicit Drugs and Transnational Criminal Threats. The strategy does not fix responsibility for a lead agency for the CTOC war, but it provides that the ONDCP-led Interagency Policy Committee will "issue implementation guidance, establish performance measures, and receive regular progress updates from the interagency community."[26]

The implication for SOF is they must continue to train for meeting the strategic challenges ahead. This will require forward-deployed units that are engaged with their counterparts in the host countries because TOC is both a threat to and a result of weak, emerging democratic governments that need help. Readiness to conduct all the SOF core activities will remain a priority, with FID, security force assistance, counterinsurgency, CT, and unconventional warfare deserving special attention.

The Findings

In this report of the proceedings, the authors have provided their analysis of a broad range of issues related to TOC. The chapters in this report are penned by practitioners who are active in operations, policy, and research concerning crime and terrorism. Readers will want to compare the various views about the nature of the threat and what can be done about it. Most important is the authors' findings on the SOF role in combating TOC. The chapters have been organized here as they relate to the strategic environment, policy and strategy, and operational issues.

The Strategic Environment

Transnational Organized Crime, a Regional Perspective

Brigadier General (retired) Hector E. Pagan takes a close look at the current environment of transnational crime and corruption throughout Central and South America. While Douglas Farah finds that the overlap of state corruption, terrorists, and TOC "applies most particularly to the Bolivarian states," General Pagan's chapter shows the problem is widespread in the region.[27] He identifies TOC as the greatest threat to the nations of the hemisphere. This is particularly evident in the high rate of crime in many of the major cities of the region. Violence between drug cartels and gang violence is the main cause for the uptick in murders.

General Pagan alerts the reader that while TOC continues to increase in the Western Hemisphere, U.S. aid for the region is steadily decreasing. As a result, SOF engagement activities there have decreased by a third. Alas the budget history shows that funding fluctuates as Congress vacillates over the urgency of the terrorism threat. "When the terrorist threat seems less prominent in the perception of budget officers, funding for overseas CT programs typically assumes a lower priority."[28]

General Pagan gives CTOC efforts in the region a high priority. He enjoins the United States, Canada, Mexico, and other regional nations to integrate efforts to establish a "defense in depth" to counter TOC. He provides 15 objectives for an integrated regional approach by the United States and its neighbors that the reader will want to consider.

SOF and the New Borderless World

Dr. Emily Spencer argues that "the predilection to isolationism is now anachronistic" for countries that must function in an environment of globalization and a virtual borderless world. "[W]hat happens 'over there' more often than not impacts us 'over here.'" Failed and weak states, potent non-state actors, violent extremist groups, and transnational criminal organizations define our surroundings in which today's threats are "exponentially greater than those of the Cold War era."

Dr. Spencer advises that in such a world, SOF have become an important tool for Western governments to meet the security challenges in a borderless world. SOF have substantive capability to deal with the threat: high readiness, credibility in sensitive domains, a low profile, and specialized training to deal with unconventional situations.

In this regard, CANSOF has become the "force of choice" for Canada when dealing with complex situations of national import. Dr. Spencer recommends that we tackle issues at their root, namely 'over there,' and that like-minded nations with common interests should share collaborative networks to meet the threats. In so doing, SOF are especially well suited to help nations meet their responsibilities for stability and security.

Transnational Organized Crime in an Era of Accelerating Change

Mark Hanna describes how TOC contributes to instability and poor governance in vulnerable countries. He relates how globalization, while contributing positively to global commerce, has also facilitated the activities of criminal organizations. A particularly problematic result of this is the linkage between the lucrative drug trafficking business and terrorist groups.

Especially interesting is Mr. Hanna's description of the rise of criminal service providers, or the crime-as-a-service business model, that facilitates TOC. "These service providers are typically not aligned with a single organization, but perform services for a variety of illicit actors," according to Mr. Hanna. Interestingly, the crime-as-a-service business model also comes with some vulnerability for TOC. Criminal groups who use the same service providers (e.g., money launderers, weapons smugglers) "can produce leads against multiple criminal organizations" for law enforcement.

But Mr. Hanna also finds that traditional, well-structured TOC groups that are regionally based are still dangerous. They threaten weak states and tend to be the most violent groups because they must sustain and protect their operations. He concludes that increased public/private partnerships are necessary to tackle TOC since "Private companies are on the front lines of many criminal attacks and have access to important information about organized criminal groups, trends in criminal activity, and new techniques used by criminal actors."

Transnational Criminal Organizations and the SOF Nexus

Brigadier General Mike Rouleau suggests that an integrated regional approach is required to counter TCOs. TCOs have the money to corrupt the institutions of government and society; this can erode state legitimacy as political structures and security are undermined. As more people move from villages to the cities, the task of countering TCOs that operate from the sanctuary of poorly governed urban zones will become more difficult. And as the Islamic State of Iraq and Syria (ISIS) and the like challenge Western nations from the sanctuary of acquired territories, those democratic countries are reluctant to intercede in overseas areas to beat back the threat.

To win in this complex, indeed chaotic environment, General Rouleau identifies five factors requisite for a SOF response to the TOC threat. Throughout the chapter he stresses the necessity of a joint-combined-interagency approach, a Global SOF Network to pull together planning and operations, the need for agility and flexibility in environments of chaos and complexity, and the need to have a small footprint in host countries as they take the lead in CTOC operations. General Rouleau concludes that success will be achieved by SOF that have earned the credibility and trust among host nations and civil institutions needed to outpace TOC.

Transnational Organized Crime: The SOF Nexus

Colonel Earl Vandahl addresses the nexus of SOF and TOC. Colonel Vandahl's extensive experience with Canada's Strategic Joint Staff enables him to share important insights about how SOF might fill the gaps and enhance the effectiveness of inter-ministerial networking that is critical for a coordinated national approach to countering TOC.

He finds that high readiness and small operating footprint of SOF makes them well suited to deal with TOC issues overseas. This is especially true of applying indirect strategies that seek to enhance host nation capabilities so local forces are trained to deal effectively with TOC. In advancing coordinated national approaches to TOC, Colonel Vandahl cautions readers about the inherent difficulty of achieving interagency cooperation if only for the diversity of core mandates of the various agencies involved in countering TOC.

Colonel Vandahl's assertion that SOF can be organized to fight as an extension of domestic law enforcement for both overseas and rural domestic settings may give pause to the reader concerned about Constitutional and regulatory proscriptions concerning the use of military forces within the homeland. But he argues that the specialized capabilities that SOF offer, particularly intelligence gathering, will be especially helpful in integrated SOF and law enforcement operations to defeat TOC. Special permissions, such as Presidential Findings and newly enacted laws, can make SOF and law enforcement interaction permissible even in a domestic setting, making Colonel Vandahl's concepts practicable. Colonel Vandahl concludes that SOF are particularly well positioned to employ specific attributes and capabilities in the fight against TOC.

Policy and Strategy

CTOC Strategic Guidance

Dr. Pete McCabe discusses the United States, Canada, and Mexico policy and strategies and how they determine to counter TOC. He lays out policy similarities and differences and searches for the guidelines needed to engage SOF in the CTOC effort. Alas, CTOC policy in North America is lacking, but Dr. McCabe does find niches in which SOF unique capabilities might fit (e.g., assisting weak states, supporting law enforcement), particularly in Mexico where the military plays a big role in supporting law enforcement.

The reader will find especially interesting Dr. McCabe's extensive analysis of Mexico's security environment and its National Security Program. Dr. McCabe rightly concludes that the CTOC challenges are daunting. The borders north and south remain porous, as are ports of entry. Drug trafficking threatens the three countries, causing grave domestic security issues, particularly in Mexico. This is critical because many trafficking groups are associated with terrorist organizations. While learning from Mexico's experiences, U.S. and Canadian SOF can assist Mexico in achieving their security goals. With law enforcement having the lead in CTOC, the SOF role will continue as support, train, and assist.

North American Efforts to Combat the Financing of Terrorism

Professor Celina Realuyo provides a comparison of the Canadian, Mexican, and U.S. efforts to counter threat finance (CTF). She underscores the importance of focusing our efforts on combating threat financing because criminals and terrorists need financing to enable their operations to succeed.

And knowing how illicit groups are financed "is instrumental in devising strategies to counter and neutralize them."

Professor Realuyo describes Canada's regime of policy and legislative permissions that arm the Royal Canadian Mounted Police, Canadian Security Intelligence Service, Financial Transaction Reports Analysis Center, and other agencies with measures that can disrupt terrorist threats and money laundering. The Mexican government has also "strengthened its legal framework to address … threat financing." The operations of Mexican agencies like the Financial Intelligence Unit have benefited from amendments to the Federal Penal Code and anti-money laundering legislation.

In the United States, financial intelligence is a critical part of the CTOC and CT efforts, and Realuyo identifies 22 U.S. agencies that are involved in this effort. Since September 2011, the United States has applied the financial instrument of national power to follow the money; and Professor Realuyo sees this as essential for detecting, disrupting, and dismantling criminal and terrorist groups.

Inside Pandora's Box: Foreign Fighters and the Lone Wolf Terrorism Nexus

Colonel Bernd Horn's chapter reviews another dimension of transnational criminal activity: the foreign fighter and related lone wolf mode of attack. Colonel Horn makes a strong and well-documented argument of the seriousness of the foreign fighter/lone wolf threat to domestic society.

In his chapter, Colonel Horn warns that volunteer foreign fighters, "individuals who leave their home country to participate in conflict in another state … expand the international reach" of transnational threats. His research reveals that about 40 percent of the ranks of ISIS are filled by foreign fighters, and about 10 percent of foreign fighters who return to their home countries become involved in domestic terrorist plots.

In discussing the "blow-back effect" Colonel Horn addresses individual motives and vulnerabilities to increase our understanding of the challenge. To meet this challenge, Colonel Horn describes Canada's Combating Terrorism Act and how the country plans to counteract foreign fighter and lone wolf attacks.

Operations

Borders and Security

Dr. Christian Leuprecht describes the effectiveness of transnational criminal networks in penetrating traditional barriers delineated by international borders as well as the opportunities these networks can provide for law enforcement to disrupt TOC. International frontiers, protected by government laws and policy, remain important for countries to maintain their sovereign domains *uti possidetis juris*. Yet deviant globalization effectively leverages legal barriers to meet the "demand for goods and services that are illegal or considered repugnant in one place by using a supply from elsewhere in the world where morals are different or law enforcement is less effective.

Dr. Leuprecht explains how networks have become a force multiplier for TOC. He walks the reader through some of the basics of network science and the associated metrics of network structure, such as centrality measures. Social network analysis (SNA), Dr. Leuprecht advises, is critically important for analyzing and disrupting TOC. He asserts, "Networks are the most important unit of analysis in understanding the formation and dynamics of illicit organizations today."

The premise of this chapter is supported by Dr. Leuprecht's examination of a counterdrug operation conducted by police in Ontario, Canada, called Project Corral. Through the use of SNA, police were able to identify key personnel in the Canadian and Jamaican drug trafficking organizations. Dr. Leuprecht concludes, "evidence suggests that [TOC] structures are both flatter and more autonomous than generally assumed," and more importantly, "borders, networks, and their nexus matter to detecting, dismantling, and deterring organized illicit activity."

An Ontological Framework for Understanding the Terror-Crime Nexus

Colonel Bill Mandrick provides the reader an ontological framework for understanding the nexus between terrorism and TOC, and organizing information about this phenomenon.[29] This framework can provide an understanding of the kinds and structures of objects, properties, events, processes, and relations in every area of reality as they relate to TCOs. The timely and accurate organization of this data creates information that is the lifeblood of Joint Intelligence Preparation of the Operational Environment (JIPOE).

In his chapter, Colonel Mandrick applies the tools of philosophical ontology to the nexus between two domains—terrorism and TOC. The prescribed ontological framework facilitates the creation of nodes and links, which represent the constituent elements of these two domains. This data feeds into planning, targeting, and JIPOE.

Thoughts on Special Operations Forces Roles in Combating Transnational Organized Crime

Mr. Randy Paul Pearson investigates the use of SOF-unique capabilities that can be effective in countering TOC. He compares the old days when the likes of Al Capone ruled well-structured criminal organizations to the environment of today where organized crime involves a loose amalgam of sophisticated, networked groups that may constantly shift interests and allegiances. While decapitation strategies can be helpful in disrupting TOC, as it was against the Capone organization, this approach does not necessarily lead to dismantlement.

SOF analysis methodology, Mr. Pearson says, can be useful to interagency departments and civilian institutions as they examine today's complex domestic and international environments and the criminal organizations they enfold. The discipline of determining criticality, accessibility, recuperability, vulnerability, effect, and recognizability, with regard to TOC, could prove to be the key to identifying the most effective course of action and the elements of national power needed to dismantle criminal organizations.

Mr. Pearson concludes that SOF's role in countering TOC could be to help with interagency communications, analysis, and information sharing. "SOF can help build a whole-of-government information sharing enterprise that can be shared across all levels of government and law enforcement, possibly developing a new way of orienting all elements of national power to countering the TOC threat."

Conclusion: SOF Roles and Future Challenges

World Order or Disorder: the SOF Contribution

Mr. Michael Miklaucic concludes these symposium findings with an examination of the challenges facing states to their sovereign place in the international political domain. He finds that the Westphalian notion of a rule-based system is being gravely challenged by the convergence of illicit networks of terrorists, insurgents, and transnational criminals.

Weak and failing states are especially pressed to maintain sovereign control of the use of legitimate coercive power and provide for the common good. He writes, "Cartels and gangs, as well as terrorists, and some insurgents, can now outman, outspend, and outgun the formal governments of the countries where they reside."

Mr. Miklaucic finds that it is imperative that efforts be made to meet the need for state-building to counter the scourge of converging illicit networks. While this will require the application of political, economic, and security resources that are in short supply, it is critical that the globe's viable states make efforts to help ailing countries that are under pressure from various armed groups and TOC.

Mr. Miklaucic finds that SOF "have unique capabilities and strengths that can help meet the sovereignty challenges facing our partner states." They can both support military and law enforcement operations and training while also participating in non-kinetic activities such as civil affairs—building partnerships, medical and veterinary aid, and setting standards of conduct for host nation militaries to emulate. In these ways, SOF can serve as a major factor in strengthening weak states. Forward looking, Mr. Miklaucic asserts today's illicit transnational networks threaten our system of nation-states, and these networks "must be countered if the system is to survive the current generation intact."

Endnotes

1. Barack Obama, *Strategy to Combat Transnational Organized Crime* (Washington, D.C., The White House, July 2011).
2. United Nations, General Assembly, A/RES/55/25, *United Nations Convention Against Transnational Organized Crime*, 8 January 2001, 4.

3. Douglas Farah, "Transnational Organized Crime, Terrorism, And Criminalized States In Latin America: An Emerging Tier-One National Security Threat," (Carlisle, PA: Strategic Studies Institute, 2012), viii.
4. Council on Foreign Relations, "The Global Regime for Transnational Organized Crime," accessed at: http://www.cfr.org/transnational-crime/global-regime-transnational-crime/p28656. The top 10 in terms of global black market value in billions are: counterfeit drugs, $200; prostitution, $186; counterfeit electronics, $169; marijuana, $42; gambling, $140; cocaine, $85; prescription drug abuse, $73; heroin, $68; software, $63; cigarettes, $50, accessed 14 July 2015 at: http://www.havocscope.com/tag/counterfeit-drugs/.
5. UN Office on Drugs and Crime, Money Laundering, "Transnational Organized Crime," http://www.unodc.org/toc/en/crimes/organized-crime.html accessed 14 July 2015.
6. UN Office on Drugs and Crime, Money Laundering, accessed at: http://www.unodc.org/unodc/en/money-laundering/globalization.html.
7. William F. Wechsler, Deputy Assistant Secretary of Defense for Counternarcotics and Global Threats, remarks prepared for delivery at The Washington Institute 26 April 2012, 3. Accessed 14 July 2015 at: http://www.washingtoninstitute.org/html/pdf/WechslerPrepared20120426.pdf.
8. "The Big Guerrilla Business," *Semana*, 7-14 July 1992, 26-32.
9. Wechsler, The Washington Institute remarks, 5.
10. Michael F. Walther, "Insanity: Four Decades of U.S. Counterdrug Strategy," Carlisle Papers (Carlisle, PA: Strategic Studies Institute, 2012),17. Here, Colonel Walther quotes from Albert Einstein, "My First Impressions of the U.S.A.," in *Ideas and Opinions* (New York: Three Rivers, 1954): Einstein comments about prohibition law, "nothing is more destructive of respect for the government and the law of the land than passing laws which cannot be enforced. It is an open secret that the dangerous increase of crime in this country is closely connected with this."
11. George Bush, *National Drug Control Strategy*, (Washington: The White House, 5 September 1989), 9.
12. Ibid.
13. William J. Perry, "Memorandum for Secretaries of the Military, Subject: Department of Defense Guidance for Implementation of National Drug Control Policy," 27 October 1993.
14. JTF-4 is now named JTF South under U.S. Southern Command; JTF-5 is now JTF West under Pacific Command at Camp H.M. Smith in Hawaii.
15. Joint Task Force Six, Command Briefing, El Paso, Texas, June 1990.
16. "Joint Task Force North supports Drug Law Enforcement Agencies in the conduct of Counter Drug/Counter Narco-Terrorism operations in the USNORTHCOM theater of operation to disrupt trans-national criminal organizations and deter

their freedom of action in order to protect the homeland," accessed at: http://www.jtfn.northcom.mil/subpages/mission.html.

17. Barack Obama, *National Security Strategy* (Washington, D.C., The White House, February 2015), 2.
18. Ibid., 10.
19. USSOCOM Publication 1, Doctrine for Special Operations, 5 August 2011, 20. A discussion of core operations and activities are provided herein.
20. Obama, *National Security Strategy*, letter of introduction, 2.
21. Barack Obama, *Strategy to Combat Transnational Organized Crime* (Washington, D.C., The White House, July 2011), 23.
22. U.S. Southern Command, SOUTHCOM Component Commands & Units. "JIATF South is an interagency task force that serves as the catalyst for integrated and synchronized interagency counter-drug operations and is responsible for the detection and monitoring of suspect air and maritime drug activity in the Caribbean Sea, Gulf of Mexico, and the eastern Pacific. JIATF-South also collects, processes, and disseminates counter-drug information for interagency operations," accessed at: http://www.southcom.mil/aboutus/Pages/Our-Team.aspx.
23. Obama, *National Security Strategy*, 27.
24. Joint Task Force North, Mission. Joint Task Force North supports Drug Law Enforcement Agencies in the conduct of Counter Drug/Counter Narco-Terrorism operations in the USNORTHCOM theater of operation to disrupt trans-national criminal organizations and deter their freedom of action in order to protect the homeland. Accessed at: http://www.jtfn.northcom.mil/subpages/mission.html.
25. Thomas Kuhn, *The Structure of Scientific Revolutions* (Chicago: University of Chicago Press, 1996), 96. Kuhn reminds that history repeats as we select familiar and comfortable pathways: "Normal research, which is cumulative, owes its success to… [selecting] problems that can be solved with conceptual and instrumental techniques close to those already in existence."
26. Obama, *National Security Strategy*, 4.
27. Farah, "Transnational Organized Crime, Terrorism, And Criminalized States In Latin America," 48. Farah writes: "The self-proclaimed "Bolivarian" states (Venezuela, Ecuador, Bolivia, and Nicaragua) take their name from Simón Bolivar, the revered 19th century leader of South American independence from Spain. They espouse 21st century socialism, a vague notion that is deeply hostile to free market reforms, to the United States as an imperial power, and toward traditional liberal democratic concepts," 68.
28. Michael B. Kraft and Celina B. Realuyo, "U.S. Interagency Efforts to Combat International Terrorism Through Foreign Capacity Building Programs," *Case Studies Working Group Report*, Project on National Security Reform, Vol II, Richard Weitz, ed. (Carlisle, PA: Strategic Studies Institute, March 2012.), 667.

29. Ontology as a branch of philosophy is the science of what is, of the kinds and structures of objects, properties, events, processes and relations in every area of reality. In simple terms it seeks the classification of entities in some domain of interest. Philosophical ontologists have more recently begun to concern themselves not only with the world as it is studied by the sciences, but also with domains of practical activity such as law, medicine, engineering, and commerce. They seek to apply the tools of philosophical ontology in order to solve problems, which arise in these domains.

The Strategic Environment

Chapter 1. Transnational Organized Crime, a Regional Perspective[1]

Brigadier General (retired) Hector E. Pagan

As I have lived, experienced, studied, and deployed to the Latin American region for decades, like many others, I continue to look for reasons to be optimistic about the region. The sad reality is that I often come short. The cycle of drug-related crime, violence, terrorism, corruption, and illegal immigration keeps the region mired in an inescapable dilemma of countries using scarce resources to make life better for people everywhere or maintain armies they cannot afford. TOC continues to be the highest threat in our hemisphere. The tendency has been to look at this criminal activity as a United States problem on one end and as a production problem on the other, with Colombia and Peru taking turns at the number one spot as the top cocaine producer in the world.

First, we need to review the current regional environment. In spite of decades of suffering through the effects of drug cartel activity and becoming the biggest consumer of illicit drugs, I believe that in the United States we don't view this as big a problem as it should be. It is understandable that al-Qaeda, the Taliban, and now ISIS dominate the news. However, we have diminished our attention to the transnational crime problem at our own peril. In his most recent testimony to the U.S. Congress, Marine Corps General John Kelly, United States Southern Command (SOUTHCOM)

U.S. Army Brigadier General Hector E. Pagan retired in 2011 as the Commander of the Special Operations Command South. His Special Operations career includes tours in the 7th, 1st, and 5th Special Forces Groups. He also served in the Army Special Operations Command, JFK Special Warfare Center and School, and United States Special Operations Command. He is a Distinguished Senior Fellow with the Joint Special Operations University.

commander, said the drug cartels are more efficient than FedEx. The analogy is right on the mark as the cartels operate as a business with organization, hierarchy, functional alignment, and competition that would rival any Fortune 500 company.

The evolution of a Colombian dominated drug trafficking operation to a Mexican one brought more money into the equation, increased violence, widespread corruption, and improved efficiency. In a move you would expect from a major corporation, Colombian cartels passed the responsibility for the movement of drugs from Honduras and Guatemala, through Mexico into the United States. The Colombian cartels formerly dominated the entire process but shrewdly recognized that even with reduced revenue, the lesser risk made this arrangement worthwhile. The Mexican cartels eagerly took that side of the business and became the dominant force, especially along the U.S.–Mexico border area.

According to a 2014 Rand Corporation research brief titled "How big is the U.S. Market for Illegal Drugs?" in spite of a 50 percent decline in cocaine use between 2006 and 2010, the drug business is at $100 billion annually. It must be noted that the decrease in cocaine use has been followed by an estimated 30 percent increase in marijuana use, a steep decline in methamphetamine use, and stable numbers in heroin consumption.[2] It is a business, and business is good for those living off the drug trade, especially in Mexico.

At one point, Mexican newspapers published articles announcing the death of the Sinaloa Cartel leader's chief of security. Their crime boss, Joaquin "El Chapo" Guzman was behind bars and recently escaped again. Los Zetas' top leader and others in the organization have also been arrested. Beyond the obvious importance and nature of these arrests, the real story is that if you read in the newspapers that the number three leader in a given drug cartel has been arrested, that means that there is a number two and a number four. It means there is an organization, orders are passed, and someone is in charge of warehousing, distribution, money laundering, smuggling through the U.S.–Mexico border and ultimately distribution inside the United States. Since 2009, Mexican authorities have captured or killed two-thirds of their most wanted drug traffickers with a substantial disruption of the flow of drugs. However, these groups have proven to be capable of regenerating leadership spots and running operations from jail. In a typical case of no decision without risk or action without reaction, the arrests brought unprecedented levels of violence. No good deed goes unpunished.

Violence follows transnational crime. The top 10 most violent cities in the world are in Latin America. This dubious distinction belongs to San Pedro Sula in Honduras (171.2 homicides per 100,000 residents); followed by Caracas, Venezuela; Acapulco, Mexico; Joao Pessoa, Brazil; Central District, Honduras; Maceio, Brazil; Valencia, Venezuela; Fortaleza, Brazil; Cali, Colombia; and Sao Luis, Brazil. Wars in Afghanistan and Iraq and widespread violence in Africa make this top 10 list hard to believe. Business Insider reports that one-third of global homicides occur in Latin America, even though only 8 percent of the world population lives in the region.[3] For example, in El Salvador, in March 2015, 481 people were murdered—16 people per day—a 52 percent increase from the same period last year.

The question is: who is behind the violence? According to BBC News, most of the violence in Mexico is between rival gangs.[4] The objective is control over territory and routes. The most notable are Los Zetas and the Sinaloa Cartel. Los Zetas operate in over half of the country and with much more regional control; they have become the leaders in violence. Vigilante groups and government attempts to counter these organizations have also added to the level of violence. Violence becomes a language, and the message is that they run everything in their area and people who don't comply must move, or worse yet, pay the consequences.

Corruption is a key factor in this cycle of drug-related crime, violence, terrorism, and illegal immigration. The collusion of drug cartels, local governments, federal government officials, and the powerful elite, and in some cases law enforcement agencies and armed forces, is well documented. Mexico's President Peña Nieto said: "Today there is, without a doubt, a sensation of incredulity and distrust ... there has been a loss of confidence and this has sown suspicion and doubt."[5] The World Justice Project's Rule of Law Index for 2014 placed Mexico 79 out of 99 in close proximity to Russia, Madagascar, China, and Egypt, in strength of rule of law, quality of governance, and criminal justice system. In Latin America it is only behind Guatemala, Nicaragua, Bolivia, and Venezuela.[6]

Mexican Armed Forces have taken a lead role in fighting criminal elements as law enforcement has fallen short of the mark. Vigilante groups have emerged in places where lack of government presence or inaction forced them to take the law into their hands. The recent disappearance in the state of Guerrero of 43 Mexican students, presumed dead, highlights this problem. In spite of an investigation that produced multiple arrests, confessions, sworn testimonies, and expert reports, there hasn't been closure on this heinous crime. Protests in the capital, Mexico City, demand answers in another scandal that has rocked the current government, undermines its credibility, and erodes confidence in the legal system.[7]

Recently, the terrorism–drug trafficking nexus has been the subject of numerous articles and congressional testimony. We have known for some time that the same routes used to smuggle drugs can be used to transport just about anything. General Kelly has expressed concern about Iran's presence in Latin America and about the radicalization of Muslims in the region. Hezbollah has expanded its operations in Latin America and into the United States and is generating millions of dollars through drug trafficking and money laundering, as discussed recently at a National Defense University conference. This link certainly includes Mexican drug cartels. In 2014 the U.S. Border Patrol apprehended 474 aliens from terrorist-linked countries attempting to enter the U.S. illegally.[8] General Kelly also stated concerns about Islamic extremists traveling to Syria to participate in jihad, and last year 19 Trinidadian Muslims were detained in Venezuela for conducting training with high-powered weapons.[9]

What are we doing in the region? According to the Congressional Research Service, for fiscal year (FY) 15, the $1.3 billion aid budget request for Latin America continues a downward trend. It is 10 percent lower than FY14 and 27 percent lower than FY12. The U.S. stated priorities of promoting economic and social opportunity, ensuring citizen security, strengthening effective institutions of democratic governance, and securing a clean energy future do not address the issue of TOC adequately. Therefore, any plans to fight drug cartels and criminal groups will certainly lack the resources needed. Historically, Colombia, Peru, El Salvador, Guatemala, and Honduras have been priorities. Since 2008, Mexico, through the Merida initiative, has received aid to fight crime and drug trafficking. The FY15 budget included $137 million for this initiative, also showing a decrease in funding for much needed help to fight transnational crime, secure the border, and address other programs.[10]

United States Special Operations Forces (USSOF) engagement with Latin America has seen a significant decrease since the beginning of operations in Iraq and Afghanistan. These operations took away SOF units that otherwise would carry the bulk of the engagement activities in Latin America, in support of SOUTHCOM. The Special Operations Command South saw its pool of available resources go down to about a third of what it used to have. Under this paradigm, it became increasingly difficult to maintain enduring relationships with our partners in the region and in most cases reduced engagement activities to episodic events or having to move SOF units from one country to another within a deployment cycle. Regionally, countries like Colombia and Mexico are capable of providing assistance and expertise to their neighbors, but this is not yet working at its fullest potential.

As we look at these factors we must also consider the state of poverty in the region. According to the World Bank, one in five people in Latin America, around 130 million, live in poverty and have never known any other way to live. This means, by definition, people who live on less than $4.00 a day.[11] Undeniably, poverty creates the conditions for corruption and crime, placing underserved populations in an inescapable cycle of despair. Lack of security and government services create a void that criminal groups rush to fill. Once these criminal elements replace the government they become the law. Examples are the favelas in Brazil. Gangs emerge from poor neighborhoods and in many places, primarily in Honduras and El Salvador, they have become the face of the drug traffic. However, this is not just a Central American problem. Beyond the notorious Mara Salvatrucha and Barrio 18, in Colombia's 6 biggest cities—Medellin, Cali, Bucaramanga, Barranquilla, Cartagena, and Bogota—there are 500 street gangs. In Bogota alone there are 107.[12]

What should we do? We must have a regional strategy that brings us all together to fight a common threat. The United States, Canada, and Mexico have to look at transnational crime as a regional problem. It is time, again, to look at this as a defense in depth. The strategic end state could be a Latin American region capable of disrupting the flow of illegal trafficking of all kinds, with a network of partner nations sharing information and intelligence, establishing a defense in depth from production to distribution.

We must understand that this is an enterprise, albeit criminal, and approach it as such. A coordinated, comprehensive plan should focus on the entire flow: production, movement, storage, smuggling, distribution, overseas network, and money laundering. Those who disregard these efforts fail to realize that we have never had the proper resources available to find out how well this might work. This whole-of-government, regional approach will work if we did some of the following as a minimum:

U.S. Defense Department and Partner Nation Ministries

1. Invest to train, equip, follow-up, and then do it again.
2. Work together and assist with planning efforts and strategy development.
3. Build the regional network. Bilateral relations with the U.S. are good but not enough.
4. Improve coordination and communication among neighboring nations. Who do you call to pass intelligence?
5. Review and agree on how and where to do interdiction operations, especially in the source zones.

U.S. State Department and Partner Nation Ministries

1. Encourage, promote, and facilitate investment in Latin America.
2. Remove trade barriers and establish commerce and technological cooperation zones.
3. Spend money in Latin America—everyone will benefit from this.
4. Review and implement flexible authorities with resources to match so it doesn't take an inordinate amount of time to execute projects or purchase needed equipment.

U.S. Law Enforcement Agencies and Partner Nation Ministries

1. Develop strong, mutually supportive agreements with SOF and other military units capable of supporting interdiction operations.
2. Resource law enforcement agencies' mission at embassies adequately with the required skills.
3. Embed with partner nation agencies regionally.

The North American Region

1. Focus on Central and South America and the Caribbean because our collective future depends on it.

2. If we do the above then securing the borders would be easier.

The Western Hemisphere

1. Set aside differences, concentrate on regional goals, and focus on the common threats.

These extreme circumstances place all of us in a position of having to do something about transnational crime, its causes, actors, and factors. No country can do it alone, as it is not a problem that belongs to any country in particular. We have much to do and it must start by seeking ways to work together. If we cannot effectively work together, the criminal cycle that ties together drug traffic, terrorism, corruption, money laundering, and violence will remain a threat to our way of life, regional stability, and ultimately safety in our streets and neighborhoods everywhere.

Endnotes

1. This chapter is based on remarks given at the "SOF Role in Combatting Transnational Organized Crime" Symposium, 8-9 April 2015.
2. Rand, "How Big is the U.S. Market for Illegal Drugs?," 2, accessed 10 July 2015 at: http://www.rand.org/content/dam/rand/pubs/research_briefs/RB9700/RB9770/RAND_RB9770.pdf.
3. Amanda Macias and Pamela Engel, "The 50 Most Violent Cities in the World," *Business Insider*, 23 January 2015, accessed 10 July 2015 at: http://www.businessinsider.com/the-50-most-violent-cities-in-the-world-2015-1.
4. "Who is behind Mexico's drug-related violence?", BBC News, 10 February 2014, accessed at http://www.bbc.com/news/world-latin-america-10681249.
5. Nathaniel Parish Flannery, "Can Mexico's Peña Nieto Walk the Walk Against Corruption?", *World Politics Review*, 5 March 2015, accessed 9 July 2015 at: http://www.worldpoliticsreview.com/articles/15221/can-mexico-s-pena-nieto-walk-the-walk-against-corruption.
6. World Justice Project, *Rule of Law Index*, 174, accessed 9 July 2015 at: http://worldjusticeproject.org/sites/default/files/files/wjp_rule_of_law_index_2014_report.pdf.
7. Ryan Devereaux, "Ghosts of Iguala," *First Look Media*, 4 May 2015, accessed at: https://firstlook.org/theintercept/2015/05/04/mexico-ayotzinapa-43-students-disappeared-part-2/. See also Elizabeth Malkin, "Suspect Held in Disappearance of 43 Mexican Students," 16 January 2015, accessed 9 July 2015 at: http://www.nytimes.com/2015/01/17/world/americas/suspect-in-disappearance-of-43-mexican-students-is-held.html?_r=0m.

8. Edwin Mora, "474 Illegals from Terrorism-Linked Countries Apprehended in 2014 Alone," Breitbart, 19 September 2014, accessed 14 July 2015 at: http://www.breitbart.com/national-security/2014/09/19/474-aliens-from-terrorism-linked-countries-apprehended-in-2014-alone/.

9. General John P. Kelly, USMC, Posture Statement of U.S. Southern Command, 12 March 2015, 7, accessed at: http://www.southcom.mil/newsroom/Documents/SOUTHCOM_POSTURE_STATEMENT_FINAL_2015.pdf.

10. Clare Ribando Seelke and Kristin Finklea, "U.S.-Mexican Security Cooperation: The Mérida Initiative and Beyond, Congressional Research Service, 7 May 2015, Table A-1, accessed at: https://www.fas.org/sgp/crs/row/R41349.pdf.

11. The World Bank, "Breaking the Cycle of Chronic Poverty in Latin America and the Caribbean," 9 March 2015, accessed 9 July 2015 at: http://www.worldbank.org/en/news/press-release/2015/03/09/breaking-the-cycle-of-chronic-poverty-in-latin-america-and-the-caribbean.

12. James Bargent, "Over 500 Street Gangs Terrorize Colombia's Biggest Cities," *In Sight Crime*, 7 April 2015, accessed 9 July 2015 at: http://www.insightcrime.org/news-briefs/over-500-street-gangs-terrorize-colombia-s-biggest-cities.

Chapter 2. SOF and the New Borderless World

Dr. Emily Spencer

For commanders privy to the strategic-level decisions and problem sets that exist, it is clear that we live in a complex world that is fraught with conflict and social and political issues to which there are no easy or clear solutions. Notably, the ability to walk away from these hot spots is not an option for any nation. As such, as a global citizen, Canada must carry its share of the burden of establishing and maintaining global security and stability. Even if the Canadian public does not always see a need to act, if Canada is to maintain influence in the world and with her partners, particularly her trading partners, then the country must be seen as being a contributing member to global stability. With challenging threats to Canada's national interests such as terrorism, narco-trafficking, and TOC, the predilection to isolationism is now anachronistic. Globalization and increased media technologies have created a virtual borderless world. Events in one part of the world reverberate around the globe. In the end, no nation, including Canada, is immune from threats emanating in seemingly far off places.

In retrospect, the Cold War was a relatively simple time when the rules of international politics and conflict seemed understandable and were generally respected. After all, the bi-polar world divided between the United States and its North Atlantic Treaty Organization (NATO) alliance, and their adversary the Soviet Union and its Warsaw Pact alliance made conflict manageable. Both sides had an understanding of what was at stake, respected spheres of influence, and used proxy forces to contest areas of contention while always being careful to avoid any direct conflict between the two superpowers. Military forces were easily recognizable and leaders and politicians followed a generally agreed upon protocol. Quite simply, we

---

Dr. Emily Spencer has a Ph.D. in War Studies from the Royal Military College of Canada. She is currently the Director of Research and Education at the CANSOFCOM Professional Development Centre, and an Adjunct Professor at Norwich University and the University of Calgary.

knew who the enemies were, how they would fight, where they would fight, and how we could negotiate to avoid any actual confrontation. There was a common, even if unwritten, agreement on the rules of how international affairs and conflict would be undertaken and managed.

The 21st century world is a much more complex, dynamic, and dangerous place. During the Cold War the specter of nuclear Armageddon was always present; however, the likelihood of rational state actors actually pulling the trigger was minimal. Today, with increased globalization, technological advancements, the rise of powerful and influential non-state actors, disintegrating failed and failing states, the advance of violent extremist groups, and accompanying TOC, the threats we face are exponentially greater than those of the Cold War era.

In essence, we now live in a virtual borderless world, and the means to secure national and international security and stability increasingly rely on the ability of governments to cooperatively employ agile, high-readiness forces that can deliver precision strategic effect rapidly at minimal cost. It is for this reason SOF have become a key component in most Western governments' military tool-boxes. Moreover, when employed cooperatively as part of a Global SOF Network, they offer a potential solution to borderless threats.

The transformation of the contemporary security environment from the Cold War to where we are today did not happen overnight. Rather it marks an evolution of world affairs, scientific advancements, and conflict. In the aftermath of the fall of the Berlin Wall in late November 1989, the world underwent a number of key vicissitudes. First, it was the end of the bi-polar world. The U.S. became the sole superpower, and NATO became a powerful influence in the world. This change had a number of effects. Countries and regions that had fallen under the sponsorship of one or the other of the superpower-block alliances suddenly found themselves without essential subsidies and support. As a result, failed and failing states sprung up. With an absence of governance, generally understood to mean a loss of a government's effective control of its territory, its inability to maintain a monopoly on the legitimate use of force, the inability to provide basic public services, and the erosion of legitimate authority to make internal decisions and/or deal with the international community, some of these states spiraled into anarchy and chaos.

Without the former constraints that the Cold War had imposed, the West, and particularly NATO, apparently rejected the tenets of the 1648 Treaty of

Westphalia that had arguably become the bedrock of international relations, and which maintained that all states have sovereignty over their territory and that no external powers have a right to interfere. Instead, beginning in the 1990s, the West quickly began to intervene in the affairs of some disintegrating states, such as the Former Yugoslavia, citing the humanitarian crises and genocide as just cause. Importantly, these failing states gave rise to non-state actors, such as criminals, warlords, and terrorists who used the chaos and absence of authority to further their own ends.

This period also saw an evolution in social and political discourse. A steady diet of scandal encompassing the leadership of governments, captains of industry, religious organizations, the military, and police drove many societies from a position of trusting those in authority to one of demanding accountability and transparency of their ruling bodies.[1] As such, publics switched from having an attitude of deference to exhibiting an attitude of defiance.

Furthermore, this period witnessed exponential scientific and technological advancements in nano-science, digitization, computerization, and satellite and information technologies to name but a few. These advancements fueled globalization to an extent never before realized. For example, recent trends in globalization exacerbated identity issues by pitting the 'haves' against the 'have-nots' in intra-state, as well as regional and international contexts. Globalization also assisted in the proliferation of cheap, accessible technologies, weapons, and information. This change has empowered lesser states and non-state actors and facilitated the interaction among terrorist groups and transnational criminal groups, thus, making them more potent and resilient adversaries.

Significantly, adding to the complexity, this period also witnessed an ever-increasing ubiquitous presence of the media. Instantaneous feeds from operational areas around the globe were pumped directly into the homes of civilians around the world in real time. Beginning with CNN and other news agencies, seemingly innocuous tactical situations on the ground were catapulted into the strategic sphere. In essence, the reporting of seemingly minor events could generate hostility around the world and create international incidents for domestic governments if the actions or words of their representatives, military or civilian, were construed as disrespectful or unnecessary. This possibility became even more likely when the message was aired out of context.

The global transformation of the security environment further accelerated in the wake of the terrorist attacks on the World Trade Center Twin Towers on 11 September 2011. The resultant Global War on Terror and the corollary conflicts in Afghanistan and Iraq, fueled by the continuing evolution of societal, political, and technological structures, pushed the world closer to the virtual borderless paradigm we have today. The new, more interconnected world facilitated planning, financing, communicating, and the sharing of ideas, tactics, techniques, and procedures, as well as lessons learned for would-be antagonists. As such, non-state actors, whether transnational corporations, organized crime, narco-traffickers, or terrorist networks, to name a few, used the new technologies to push their agendas. This change made them more efficient and effective.

The explosion of the Internet and social media has added to the complexity and difficulty.[2] In essence, the world is often aware of events before the official chain of command, political and/or military, can verify or comment on the veracity of the event. Flash mobs, riots, if not entire movements, can spring up without notice due to graphics and video shared through social media. One need only remember the aftermath of the erroneous report by *Newsweek* in 2005 of U.S. servicemen flushing Qurans down the toilet at Guantanamo Bay or the 2011 calamity of Pastor Terry Jones burning a Quran and the weeks of rioting, deaths, and millions of dollars of damage that resulted in order to begin to appreciate the global reach of the Internet and social media. Perhaps most notable is the Arab Spring in 2011 that subsequently toppled a number of Middle East regimes and whose aftermath is still being felt.

In the end, we live in a very interconnected, arguably borderless, world that impacts the dissemination and conduct of information, travel, business, disease, and warfare, to name a few. Regardless of physical space, social connectivity is generally no longer delayed or deterred by geographic or political realities. As a result, physical borders are being treated as meaningless to groups that have the social strength to transcend them. In short, the borderless world of the Internet and social media has created a virtual borderless world of ideas which has given groups and movements the power to pass through the imposed geopolitical borders of the 'real world.' As such, a nation can no longer turn a blind eye to international events, since what happens 'over there' more often than not impacts us 'over here.' For instance, one need only look at the global recession, the foreign fighters issue, TOC, or the Ebola crisis to see this trend.

Conflict flourishes in this environment. It breeds humanitarian crises, which create migration and refugee issues, as well as the need for international relief efforts. For example, the Syrian Civil War itself has killed an estimated 160,000 to 190,000 people, displaced more than 6.5 million, and forced more than 3 million Syrians to flee to neighboring countries.[3] Moreover, present day conflicts often attract foreign fighters. This attraction has become a serious concern in Western nations since unprecedented numbers of Europeans and North Americans are traveling to the world's trouble-spots to fight, often for extremist organizations. The concern is particularly one of receiving a 'blow back' effect, where foreign fighters return or are sent home or to other Western countries to continue the fight. One study revealed that one in nine foreign fighters engage in an attack on their originating country or another Western nation.[4] In addition, as we have seen with the Islamic State, the conflict can very easily and quickly spread throughout the region.

The issue becomes how a nation copes with this chaotic, ambiguous, unpredictable world. There are no real rules or 'unwritten' protocols to conflict anymore. The plethora of non-state actors and rogue states make predictability and recognized processes impossible. As such, a nation requires the ability to gain rapid situational awareness to determine what exactly is occurring. Nations also need the ability to rapidly dispatch substantive capability, whether humanitarian, economic, or military, to deal with the emergency, which normally for a middle power such as Canada is as part of a coalition. Notably, the word 'substantive' is used to underscore the fact that governments require an agile, credible contribution that is recognized as such by its allies and partners. Humanitarian aid, logistical support, and/or supporting political rhetoric are not enough. To earn and maintain a seat at the table requires risk acceptance and leadership that is associated with operational capability that can and will be deployed in harm's way. This juncture is where SOF play an important role. Indeed, SOF are a critical enabler within governments' military toolboxes in this new borderless world.

SOF bring to the fore an uncontestable value proposition. First, SOF provide governments with high readiness, rapidly deployable discreet forces that can provide discriminate, precise kinetic and non-kinetic effects. These alternatives provide governments with an immediate ability to deploy capability within a wide range of options. Second, SOF are dependable and credible.

Their reliance on highly trained, technologically enabled forces that can gain access to hostile, denied, or politically sensitive areas has provided SOF with a series of operational accomplishments that have been recognized. As such, their past record and continuing performance lowers the risk of their employment, but at the same time is generally accepted as a substantive commitment of capability by allies. Also mitigating risk is the low signature/profile of SOF. Small, capable teams allow for a relative 'easy in' and 'easy out.'

In addition, SOF enhance strategic decision making. The characteristics given above support the deployment of SOF to assess and survey potential crises to provide 'ground truth' and situational awareness for government decision makers. Having accurate, precise, timely information dampens strategic risk and enables political and military decision makers to contemplate the necessary decisions and weigh the various options based on trustable up-to-date information. Finally, SOF provide a highly trained, specialized force capable of providing a response to ambiguous, asymmetric, unconventional situations that fall outside the capabilities of law enforcement agencies, the conventional military, or other government departments.

For example, CANSOFCOM, which was layered on top of JTF 2, the nation's counterterrorism unit that was established in 1993, was created to assist in meeting the requirement of the complex contemporary operating environment (COE). The germination of the command began in February 2005, when General Rick Hillier, then Chief of Defence Staff (CDS), told his general officers at a special general/flag officer seminar in Cornwall, Ontario, that, "We need an integrated Canadian Forces that consists of maritime, air, land and special forces, woven together to make a more effective military."[5] This speech marked the first time that a CDS spoke of Canada's SOF capability within the context of a fourth environment within the Canadian Armed Forces (CAF). Later that year, on 19 April 2005, General Hillier declared that he intended "on bringing JTF 2, along with all the enablers that it would need, to conduct operations successfully into one organization with one commander."[6] This decision would prove to be a major step for CANSOF. As a result, on 1 February 2006, as part of the CAF's transformation program, CANSOFCOM was created.

The purpose of CANSOFCOM was clearly articulated as the need "to force develop, generate and, where required, employ and sustain Special Operations Task Forces capable of achieving tactical, operational and strategic effects required by the Government of Canada."[7] The command consisted of a small headquarters, JTF 2, a new Tier 2 combatant unit called the Canadian Special Operations Regiment, 427 Special Operations Aviation Squadron, and the Joint Nuclear, Biological, and Chemical Defence Company, which was officially changed in September 2007 to the Canadian Joint Incident Response Unit. The command has come a long way in nearly 10 years and has proven its value to the CAF, Government of Canada, and the Canadian people as a result of its pursuit of operational excellence, quality personnel, and its agility, adaptability, and dependability in the execution of its tasks.

CANSOFCOM has demonstrated its ability on a number of short-fuse operational deployments around the globe. It has displayed the characteristics and attributes of what military analysts, political decision makers, and scholars have referred to as a "force of choice." They have done so, because CANSOF:

1. Are capable of rapid deployment into any environment;
2. Are proficient at deploying small, highly capable teams that have a low signature/are low-visibility or clandestine and do not represent a major foreign policy engagement;
3. Serve as a catalyst to unify, extend the reach, and maximize the effects of other instruments of national power;
4. Are capable of working with conventional and indigenous forces, as well as other government departments;
5. Provide the government with a wide spectrum of special operations options, lethal and non-lethal, to deter, disrupt, dislocate, and when necessary, destroy those that would do harm to the nation, its allies, and friends, or its national interests, in hostile, denied, or politically sensitive areas; and
6. Importantly, represent a highly trained and educated, adaptive, agile-thinking force capable of dealing with the threat that has not yet been identified.

Clearly, SOF are capable of providing national governments with a wide range of viable options in the face of an increasingly volatile security environment. As such, it is not difficult to imagine the power that a Global SOF Network could have. Individual costs would be lowered while simultaneously global reach would be extended and response time greatly diminished—clearly advantages in today's borderless world.

Notably, the concept of SOF networks is neither new, nor is it laden with risk. In fact, recent experience demonstrates the value of this course of action. For example, NATO SOF Headquarters was able to assure an increase in SOF partnering efforts and the expansion of overall SOF capabilities throughout the International Security Assistance Force in Afghanistan. After only six years of operations, NATO SOF Headquarters standardized SOF practices throughout Europe, resulting in an estimated fivefold increase in the number of operators deployed to Afghanistan. As such, it seems to be a natural evolution to continue to build stronger military ties with allies and partners and to extend these networks beyond NATO's purview. This increased interconnectedness is particularly relevant within the SOF community as individuals are already specially selected, trained, and equipped to perform high-value tasks within high-risk environments, including in politically sensitive or denied areas.

Specifically, we need to rally like-minded nations that share common interests and form collaborative networks of allies and partners in order to decisively engage, deny, degrade, and deter violent extremist organizations and any other menace to global stability and security. These individuals, organizations, and the perils that they represent are not deterred by national boundaries and maintain a global reach. As no nation is safe from attack, it logically follows that we are all responsible for shouldering the burden of contributing to the solution. Ultimately, strengthening SOF networks provides a collective solution to counter worldwide threats. In the end, we must accept that the new borderless world has resulted in the fact that we need to tackle issues at their root, namely 'over there' so they do not come and bite us, 'over here.' SOF are particularly adept at this task. And a Global SOF Network is even more capable than any one nation's SOF could be alone. Even as nations value sovereignty in the tradition of the Peace of Westphalia, they must recognize that today's borderless world requires borderless solutions.

Endnotes

1. Within the Western world corporate scandals such as the collapse of Enron, the continuing sexual scandals revolving around the Roman Catholic Church, the revelations of abuse of power and corruption by the judiciary, police and politicians, as well as excesses of senior military officers, all combined to create a movement for more transparency and accountability of public figures and those

in positions of authority and trust within the private sector. It led to what author and journalist Peter C. Newman entitled a move of society from deference to defiance. See Peter C. Newman, *The Canadian Revolution: From Deference to Defiance* (New York: Penguin Books, 1996).

2. In the context of this paper, the Webster's Dictionary definition of social media will be used. It provides the following definition for social media: "forms of electronic communication (as Web sites for social networking and microblogging) through which users create online communities to share information, ideas, personal messages, and other content (as videos)." Accessed on 25 May 2015 at: http://www.merriam-webster.com/dictionary/social%20media.
3. Canada, *2014 Public Report on the Terrorist Threat to Canada* (Ottawa: Public Safety Canada, 2014), 17.
4. Cited in *Jez Littlewood*, "Foreign Fighters: what to read to grasp the key issues," accessed on 28 July 2014 at: http://npsia.wordpress.com/2014/07/03/foreign-fighters-what-to-read-to-grasp-the-key-issues/.
5. Cited in Bernd Horn, "The Canadian Special Operations Forces' Legacy," in Emily Spencer (ed.), *Special Operations Forces: A National Capability* (Kingston: CDA Press, 2011), 54.
6. Ibid.
7. Canada, *CANSOFCOM Capstone Concept for Special Operations 2009* (Ottawa: DND, 2009).

Chapter 3. Transnational Organized Crime in an Era of Accelerating Change

Mr. Mark Hanna

Transnational organized crime is evolving in ways that challenge our understanding of the criminal enterprise. Globalization encourages regionally and ethnically based criminal groups to expand into international markets, provides them with access to global transportation modes and finance, and forces them to adopt more flexible organizational structures. The revolution in computer and telecommunications technologies accelerates these trends, opens new criminal markets, often at vastly larger scale, and offers greater options for collaboration among criminal groups. In addition, criminal organizations adjust their methods of operation in the face of decades of global anti-crime efforts; this is most evident with regard to drug trafficking organizations. This evolution of TOC presents challenges, as well as opportunities, for international efforts to reduce the threat posed by transnational criminal activities.

The Nature of the Transnational Organized Crime Threat

TCOs contribute to instability and undermine governance through the use of violence, intimidation, and corruption to protect and promote their criminal enterprises. Criminal groups contending for control over access to the U.S. illicit drug market are fueling high levels of homicides and corruption in Mexico and Central America. An increase in cocaine shipments from South America have resulted in high levels of corruption and instability in West Africa.[1] Criminal impunity and endemic corruption contribute to

---

Mark Hanna, the Deputy National Intelligence Officer for Transnational Threats/Narcotics, is a 30-year veteran of the Intelligence Community, having served in the National Security Agency from 1984-1990 and the Central Intelligence Agency since 1990. Most recently, he led the Crime and Narcotics Center's analytic effort. Prior to that, he served as a liaison officer with the DCI's Office of Congressional Affairs.

disenfranchisement and sense of grievance, and this is fueling popular uprisings in the Middle East.[2]

Criminal activity also provides a source of revenue for insurgencies and other non-state actors, eroding state control and at times generating threats with a global reach. This phenomenon affects many regions: the Revolutionary Armed Forces of Colombia, Sendero Luminoso in Peru, the Taliban in Afghanistan, and the Shan State in Burma are examples of insurgencies that have benefited from their access to an indigenous drug trade. Insurgents in the Congo and Nigeria fund themselves in part through the criminal exploitation of natural resources.[3]

Lastly, organized criminal activity also undermines public health, safety, and economic wellbeing. Organized crime undermines local economies by siphoning off raw materials and intellectual property, disrupts supply chains by introducing counterfeit materials, and erodes trust in the modern financial and commercial systems by fraudulently exploiting those systems. The global scale of the economic impact of illicit activity is inherently difficult to measure, but some studies estimate that the value of illicit trade is between $1 and $2 trillion dollars annually.[4]

Criminal threats are being perpetrated by an increasingly diverse set of actors, whose organizational structures and tactics evolve in response to three factors: globalization, technological advances in telecommunications and computers, and global anti-crime efforts.

Globalization

Trade liberalization has vastly increased global commerce. As a result of the opening of borders through trade liberalization, criminal organizations have access to a globally interconnected transportation network. Criminals are able to hide their illicit goods within an ever-increasing amount of licit commerce. Between 1980 and 2005, world container traffic increased tenfold, from 39 million twenty-foot equivalent units (TEU) to 395 million TEU.[5] In addition, the rise of global supply chains and just-in-time delivery of goods and services has greatly increased the pressure to keep commerce moving through ports of entry, decreasing the amount of time for inspecting goods. Commercial trade more than doubled following the passage of the North American Free Trade Agreement; according to the U.S. Customs and Border

Protection, nearly 5 million truck crossings were made from Mexico into the United States in 2007.[6]

The liberalization of global finance has also provided criminals with a vastly expanded mechanism to launder their ill-gotten gains. The UN estimates the amount of illicit funds available for money laundering would have been approximately 1 percent of global GDP in 2009, or U.S. $580 billion.[7] The global flow of illicit funds has ensnared large, international financial institutions, such as the Hong Kong and Shanghai Banking Corporation (HSBC), which agreed to pay a U.S. $1.9 billion fine as a result of a U.S. investigation that revealed money laundering in the bank's Mexico unit.[8]

The growing middle class in developing countries has fueled new markets for a variety of illicit goods such as drugs and counterfeit products, and they provide targets for a variety of fraudulent schemes. At the same time, the rise of developing markets and growing income inequality has prompted an increase in populations looking to migrate, providing a growing market for human smugglers and traffickers. The UN estimates that there were 232 million migrants in 2013, up from 154 million in 1990; most of these migrants originated in developing countries seeking to migrate to more developed economies.[9]

Technological advances

As the global economy has moved onto the Internet thanks to the computer and telecommunications revolutions, so too have criminal groups. The digital economy offers fraudsters and identity thieves access to millions of potential victims, as well as the means to market and sell stolen digital goods. The digitization of intellectual property—such as movies, music, software, and industrial design—has generated a market for goods that sophisticated cyber criminals appropriate and sell. The digitization and storage of personal identifying information has also created a burgeoning online criminal market. The Internet has globalized fraudulent schemes, giving fraudsters access to millions of potential victims. One recent estimate suggests the cost of cybercrime and cyber espionage somewhere between U.S. $70 billion and $140 billion annually; however the economic harm may be greater if one factors in such multiplier effects as the loss of reputation and competitiveness to a company's bottom-line.[10]

Criminals have also used the Internet to market illicit goods and services. Online forums provide markets for cybercrime tools such as malware and ransom-ware, as well as for skilled hackers, who can be hired for criminal services. As one Europol study notes, "The crime-as-a-service business model drives the digital underground economy by providing a wide range of commercial services that facilitate almost any type of cybercrime ... This has facilitated a move by traditional organized crime groups ... into cybercrime areas."[11] The Internet has also created a new avenue for the sale of goods such as illicit drugs, counterfeit goods, pirated intellectual property, and trafficked persons. The online marketplace 'Silk Road,' for example, offered buyers access to drugs, stolen identification cards, and hacking tools, among other illicit products.[12]

The Internet and mobile communications are also transforming how criminal actors organize themselves, fostering the development of looser organizational structures. Criminal specialists can collaborate, execute sophisticated crimes with a global reach, and disperse without physical contact. Often these criminals are working from safe havens to affect crimes in more developed and risky countries often on the other side of the globe. Moreover, the Internet has to some degree lowered the barriers to entry for aspiring criminals. They can purchase cyber tools, hire skilled specialists, and access goods and markets without having to fund and support a large organization.

Responding to global law enforcement efforts

Global anti-crime efforts have also played an important role in ongoing evolution of criminal enterprises. This is most evident when we examine the criminal organizations engaged in the global drug trade. In Mexico, for example, drug trafficking organizations have adopted flatter and more distributed organizations models after eight years of intense government pressure. In Colombia, over 20 years of antidrug efforts that spanned the hemisphere have reduced the threat from the once powerful Medellin and Cali Cartels, who have been replaced by smaller, more decentralized criminal bands that have outsourced international transportation and distribution.[13]

Environmental factors have prompted a change to the nature of criminal groups

Criminal organizations are evolving from tightly knit, ethnically, and regionally based groups to looser, more dispersed networks of criminal specialists. The fast movement of people, goods, and information stimulated by globalization and technology has encouraged transnational criminal groups to decentralize and outsource. This has led to the rise of criminal service providers such as financiers, hackers, document forgers, and transportation groups. These service providers are typically not aligned with a single organization, but perform services for a variety of illicit actors. As noted in the previous section, the environmental factors are generating a variety of organizational models: drug cartels in Mexico that are federations of smaller organizations, hub and spoke networks, and human trafficking networks that operate as chained networks of specialists (see p. 42).

These hierarchical, regionally based organized crime groups are not completely fading from the scene. In fact, the organized criminal groups that most threaten the stability in many weakly governed countries exhibit such structures. The drug trade and the criminal exploitation of natural resources in particular continue to be characterized by such groups, primarily because the manufacture and distribution of most illicit drugs requires control over territory. These groups must resort to high levels of violence, intimidation, and corruption to sustain their operations. Territorial-based criminal activity also offers the greatest opportunity for the intersection between organized groups and insurgents or other non-state actors, as previously noted in such countries as Colombia, Peru, Nigeria, and the Central African Republic.

That said, many observers believe that territorial control has become less important to modern criminal organizations than access to illicit goods, services, and markets. As a 2010 UN report on global organized crime states:

> Today, organized crime seems to be less a matter of a group of individuals who are involved in a range of illicit activities, and more a matter of a group of illicit activities in which some individuals and groups are presently involved. If these individuals are arrested and incarcerated, the activities continue, because the illicit market, and the incentives it generates, remain.[14]

Human Trafficking Networks Composed of Linked Chains

Human trafficking networks are generally comprised of small organizations that specialize in the recruitment, trafficking, or exploitation phase of trafficking that form a linked chain of organizations to move trafficked persons from source to destination countries. These networks will sometimes outsource temporary support for transportation or safe houses. For example, the Spanish took down a global human trafficking network based in China that worked with brokers in Sudan to smuggle Sudanese men into Madrid. Once in Spain, these men were enslaved in a local criminal trade to sell counterfeit goods in order to pay off the debt they owed to the smugglers.[15]

Ad Hoc Bank Fraud Network

In 2009, a transnational criminal organization exploited a sophisticated hack into a subsidiary of Royal Bank of Scotland (RBS) to withdraw more than $9 million from 2,100 ATM machines in 280 cities. The group that perpetrated the crime was comprised of an insider at RBS with access to debit card accounts, a hacker who had the technical expertise to exploit the accounts, and a third criminal who hired mules around Europe to draw millions of dollars from the accounts via ATM transactions.[16]

Online Counterfeit Drug Sales

Purveyors of counterfeit pharmaceuticals are using the Internet and cyber tools to identify and lure victims to illicit online pharmacies. Criminals with access to counterfeit drugs hire cyber specialists who build cyber infrastructure to create spam ads or affect search-engine manipulation to attract customers; build redirection sites; and develop a legitimate-looking web site to affect online purchase of counterfeit drugs. Criminals then use a variety of cash-out services to launder the online drug profits.[17]

Network Vulnerabilities Create Opportunities

Today's transnational criminal networks exhibit an increasing variety of structures that challenge our preconceptions about criminal organizations, but the globalization of criminal enterprise has also led to the weakening of the ties binding together organizations. Traditional methods used by criminals to vet collaborators are more difficult and inefficient in the global or cyber-based context. Often, relationships between criminal actors perpetrating a crime are transactional and ad hoc in nature. Traditional bonds of loyalty are limited or nonexistent. As a result, criminal organizations are in some ways easier to penetrate; criminals are more willing to give up associates.

Another potential vulnerability arises when multiple transnational criminal groups use the same criminal service provider. As a result, targeting service providers—such as money launderers, document forgers, or weapons smugglers—can produce leads against multiple criminal organizations.

The loose and flatter structures of many criminal networks can create internal stresses. Such internal stresses can be exacerbated by law enforcement efforts targeting the leadership of sub-elements within the organization. Antidrug efforts in Colombia and Mexico have fractured the larger drug cartels; although in Mexico, in particular, such efforts have provoked the inter-cartel violence as more organizations compete for the same lucrative drug market.[18]

Criminals' adoption of new technology can also make them more vulnerable. Their use of advanced telecommunications or computer technology can provide law enforcement with avenues for collecting information about individual criminals and the ability to link globally connected organizations.

Implications for Global Anti-Crime Efforts

A recent study by the Congressional Research Service summarized the challenge governments face in tackling the evolving criminal threat. It advises that criminals have expanded their range of tools and targets "while law enforcement plays by yesterday's rules and increasingly risks dealing only with the weakest criminals and easiest problems."[19] Tackling the most salient threats from TOC will require new methods for assessing risk, choosing the right targets, and developing and implementing anti-crime strategies that will require coordination across multiple jurisdictions.

Targeting Critical Players

It is often more efficacious to target specialists, such as transporters, hackers, moneymen, or chemists, rather than a criminal organization's leadership. In order to choose the actors most critical to the functioning of a criminal network, or more broadly to accurately assess the vulnerabilities of criminal enterprises, one must develop a map of the entire network. Flatter, distributed organizations are more resilient and flexible, but they can be subject to targeting efforts that weaken or eliminate critical transportation nodes, disrupt the flow of money within the organization, or weaken its security apparatus.

Disrupting Markets

Many observers now believe that targeting individuals or organizations is less effective than disrupting illicit markets, supply chains, and service providers. Disrupting supply of and demand for illicit product can reduce the incentives for organizations to stay in the game despite expensive losses, or for new players to replace dismantled organizations. Mexico's antidrug efforts have led to the arrests of numerous drug cartel leaders and the weakening of several larger criminal organizations, yet continued high demand in the United States and high levels of drug production to meet that demand continue to fuel a lucrative drug market and sustain high levels of violence between criminal organizations battling for market share. The history of efforts to limit drug supply and recent efforts to reduce the trade in ivory illustrate difficulty of disrupting illicit markets.[20]

Creating a Global Anti-Crime Network

Disrupting a globalized criminal enterprise requires a globalized response. International law enforcement efforts need to cross national boundaries with the alacrity of their criminal adversaries. However, to date, a lack of a common appreciation of the threat, competing national priorities, a patchwork of information-sharing regimes, and varying levels on institutional capacity have inhibited the development of a robust, global response to the TOC threat. And government efforts alone cannot solve the problem of TOC. Tackling modern criminal threats also necessitates increased public-private partnerships. Private companies are on the front lines of many criminal attacks and have access to important information about organized criminal groups, trends in criminal activity, and new techniques used by criminal actors. In an era of accelerating change, a global anti-crime network, strengthened by public-private partnerships, will be able to meet the challenge of TOC.

Endnotes

1. UN Office of Drugs and Crime (UNODC), *Crime and Instability: Case Studies of Transnational Threats* (New York: UNODC, February 2010), 5, 16, accessed at: https://www.unodc.org/documents/frontpage/Crime_and_instability_2010_final_low_res.pdf.
2. Stuart Levy, "Fighting Corruption After the Arab Spring," *Foreign Affairs*, 16 June 2011, accessed 23 July 2015 at: https://www.foreignaffairs.com/articles/middle-east/2011-06-16/fighting-corruption-after-arab-spring.
3. UN Office of Drugs and Crime, *Crime and Instability*, 1.
4. Justin Picard, "Can We Estimate the Global Scale and Impact of Illicit Trade?" in *Convergence: Illicit Networks and National Security in the Age of Globalization*,

Michael Miklaucic and Jacqueline Brewer, eds., (Washington, D.C.: INSS, 2013), 58. See Also, "Havoscope Black Market Ranking," and "Havoscope Black Market Country Ranking," Havoscope Global Black Market Information, 16 October 2011, available at www.havoscope.com; and UNODC "Estimating Illicit Financial Flows Resulting from Drug Trafficking and other Transnational Organized Crime," October 2011.

5. Wang, James, *Ports, Cities, and Global Supply Chains*, (Burlington, VT: Ashgate Publishing, 1 January 2007), 74.
6. John Fittelli, "North American Free Trade Agreement (NAFTA) Implementation: The Future of Commercial Trucking Across the Mexican Border," (Washington, D.C.: Congressional Research Service, 1 February 2010), accessed at: http://digitalcommons.ilr.cornell.edu/cgi/viewcontent.cgi?article=1708&context=key_workplace.
7. United Nations Office on Drugs and Crime, *Estimating Illicit Financial Flows Resulting from Drug Trafficking and other Transnational Organized Crime* (Vienna: UNDOC, October 2011), 42, accessed 23 July 2015 at: http://www.unodc.org/documents/data-and-analysis/Studies/Illicit_financial_flows_2011_web.pdf.
8. Carrick Mollenkamp, "HSBC Became Bank for Drug Cartels, Pays Big for Lapses," *Reuters*, 11 December 2012, accessed 23 July 2015 at: http://www.reuters.com/article/2012/12/12/us-hsbc-probe-idUSBRE8BA05M20121212.
9. UN Department of Economic and Social Affairs, *International Migration Report 2013*, (New York: UN, December 2013), 1, accessed 23 July 2015 at: http://www.un.org/en/development/desa/population/publications/migration/migration-report-2013.shtml.
10. James Lewis and Stewart Baker, Center for Strategic and International Studies, *The Economic Cost of Cybercrime and Cyber Espionage* (Washington, D.C.: CSIS, July 2013), 16, accessed 23 July 2015 at: http://www.mcafee.com/us/resources/reports/rp-economic-impact-cybercrime.pdf.
11. European Cybercrime Centre, European Police Office (EPO), *The Internet Organized Crime Threat Assessment*, 2014 (The Hague, Netherlands: EPO, 2014), 9, accessed 23 July 2015 at: https://www.europol.europa.eu/sites/.../europol_iocta_web.pdf.
12. Kim Zetter, "How the Feds Took Down the Silk Road Wonderland," *Wired*, 18 November 2013, accessed 23 July 2015 at: http://www.wired.com/2013/11/silk-road/.
13. Bruce Bagley, "The evolution of drug trafficking and organized crime in Latin America", *Sociologia, Problemas e Práticas*, 71-2013, accessed at: http://spp.revues.org/1010.
14. UNODC, "The globalization of Crime: A Transnational Organized Crime Threat Assessment," (Vienna: UNDOC, 2010), V, accessed 23 July 2015 at: https://www.unodc.org/documents/data-and-analysis/tocta/TOCTA_Report_2010_low_res.pdf.

15. Spanish National Police, "International human trafficking network dismantled, victims crammed into a small flat," Press Release, 1 May 2011, accessed 23 July 2015 at: http://www.policia.es/prensa/20110501_1.html.
16. Linda McGlasson, "RBS Wordplay: 8 Hackers Indicted in $9 Million ATM Theft," *Bank InfoSecurity*, 12 November 2009, accessed at: http://www.bankinfosecurity.com/rbs-worldpay-8-hackers-indicted-in-9-million-atm-theft-a-1935/op-1.
17. Nektarios Leontiadis, Tyler Moore, and Nicolas Christin, " Measuring and Analyzing Search-Redirection Attacks in the Illicit Online Prescription Drug Trade," in Proceedings of the 20th USENIX Security Symposium, San Francisco, August 2011, accessed 23 July 2015 at: http://www.contrib.andrew.cmu.edu/~nicolasc/publications/LMC-USENIXSec11.pdf.
18. Bagley, "The evolution of drug trafficking."
19. Jerome P. Bjelopera and Kristin M. Finklea, "Organized Crime: An Evolving Challenge for U.S. Law Enforcement," (Washington, D.C.: Congressional Research Service, 6 January 2012), 23, accessed at: https://www.fas.org/sgp/crs/misc/R41547.pdf.
20. Michael Sas-Rolfes, Brendan Moyle, and Daniel Stiles, "The complex policy issue of elephant ivory stockpile management," *Pachyderm*, No. 55, January-June 2014, accessed 23 July 2015 at: http://www.pachydermjournal.org/index.php/pachy/article/view/347.

Chapter 4. Transnational Criminal Organizations and the SOF Nexus

Brigadier General Mike Rouleau[1]

The COE continues to pose dynamic and evolving challenges to nations and their national security organizations. Specifically, adaptive, agile, and networked TCOs necessitate a cohesive, integrated, and regional approach in response to their insidious reach and impact. This chapter will highlight four main themes:

1. Crime and terrorism threats are converging in a globalized, borderless manner;
2. Our responses remain framed by a view of sovereign states with clear borders;
3. Since states will not disappear, we must adapt within our own imposed restrictions (i.e. jurisdictional, organizational, cultural, diplomatic); and
4. The threat and response context requires enhanced strategic authorities, but this context leads to a tactical/strategic convergence, which can push out the operational level.

TCOs, at their center, represent a convergence of threats. Quite simply, globalization has changed the way we do business (i.e. interconnected, integrated, networked, global). Not surprisingly, criminal organizations have evolved as well and have matched business trends. Importantly, they are taking advantage of advancements in business and trade to facilitate their nefarious activities. For example, "containerization," or the use of

---

Brigadier General Mike Rouleau is the Commander of CANSOFCOM. He joined Joint Task Force 2 in 1994 and served as Adjutant, Assault Troop Commander and Sabre Squadron Commander until 1999. General Rouleau assumed command of JTF 2 in May 2007, a position he held until June 2009. He also served as the Director of Special Operations Forces from 2011 to 2013, when he took command of CANSOFCOM.

sea containers, makes customs control more difficult. It is arduous enough to detect an active threat (e.g. explosive, radioactive) in one sea container among thousands, let alone benign contraband hidden in one or more sea containers.

In essence, in the manner that legitimate business seeks the best competitive advantage, TCOs do the same. For instance, Colombian drug cartels initially saw the advantage of exploiting the FARC terrorists for security. However, the FARC eventually displaced the cartels and took over the business. This takeover of the drug trade gave the FARC the opening to expand into the drug realm and exploit their integrated security, shipment, and money laundering skills. Where there is profit space, someone will try to fill it.

The important issue that arises is the fact that relatively innocuous criminal matters can expand to become greater threats. Case in point, originally cigarette smuggling across the St. Lawrence Seaway between Canada and the U.S. represented a fairly low-level menace, being more a question of denying the Government of Canada tax revenue rather than a physical threat to Canadians. However, the infrastructure, pipeline, and network quickly opened up a route that is now exploited for serious drug and weapons smuggling.

In addition, the enormous wealth of TCOs represents a further threat. Legitimate business has always sought to lobby governments to curry favor or advantage. As such, it is not surprising that criminal organizations apply the same pressures through corruption and co-opting of decision makers and local security forces. This insidious interference of governance quickly undermines legitimacy and trust in government institutions within a society, which in turn can erode or actually break down the state political and security mechanisms and processes.

In many ways, this erosion begins the destruction of sovereign states from which TCOs benefit. The more contested the level of governance and control, the greater is the freedom of maneuver for TCOs. In essence, they respect borders, which although representing boundaries, certainly do not pose as insurmountable barriers. For most nations, traditional security threats are about sovereignty and territorial integrity. From this vantage point, TCOs can appear somewhat benign since, with a few exceptions, TCOs are not interested in taking over the role of the state. While there are some instances of criminal organizations providing some local services, such as FARC and Hamas for example, they are an exception and dedicate those resources to their supporters. However, their efforts at creating 'ungoverned spaces' in which they can operate freely have a corrosive effect on the sustainability of democratic institutions.

To further add to the security challenge, there are not just fewer barriers to organized crime, but they now exist in a more complicated operating environment. Globalization and trends in the COE have also indicated the need for national security institutions to operate increasingly in urban environments to counteract the threats from TCOs. Rising global urbanization is not new. Historically people have moved from rural to urban areas in search of a better quality of life and employment opportunities.[2] Today, the developing world is awash in young men who lack sufficient legitimate job opportunities to employ them all. This situation has greatly increased criminal activity in the rings of slums that surround many of the developing world's largest cities. Not only is this environment imposing, but the threats adapt and reorganize faster than most states can respond.

Technology is a case in point. It continually reminds all of us that the state can be a lumbering beast, not optimally positioned to combat adaptive, agile TCOs. Organizational cultures must adapt and change. Two examples provide context. First, Netflix has completely changed the way people consume television, but federal authorities still seek to control broadcast distribution. The second example, Uber driving service, is completely revamping the urban transportation industry, and while municipal authorities try to crack down, it is clear they are only causing minor inconveniences to an extremely small number of Uber drivers. These are innocuous examples that are operating openly. The point is that the state's ability to respond to deliberately hidden criminal activities is certainly no better, or faster, than in previous decades.

Moreover, the difficulties of countering TCOs extend beyond borders. There are challenges eradicating the problem at its source. In the post Afghanistan/Iraq war era there is little Western public appetite or acceptance for direct intervention in other nations. The fact that governments have focused military response to the ISIS crisis in the Middle East with a careful regime of air strikes and SOF is indicative of the reluctance to become overly committed with ground forces.

Additionally, not only do Western publics not want to see a significant commitment of forces involved in these fights, but host nation governments and their publics are also reticent to see large numbers of outsiders interfering in their societies. This reality remains true even when host nation publics fully comprehend that their own political leadership is often corrupted, or at least co-opted, by criminal elements.

The challenges appear daunting, but they need not be. There are solutions and approaches that can mitigate the increasingly complex set of problems. However, what it also means is that our responses will likely need to be complex as well. I postulate that there are five factors that must be part of any SOF response to transnational crime:

1. A combined, joint, interagency approach required to bring all state power to bear;
2. The necessity for allies to build off each other's engagements in problem areas (i.e. Global SOF Network);
3. Organizational agility and flexibility in face of rapid change;
4. Locally driven solutions: Capacity building, and advise-and-assist missions should shape and assist the response, which must be host nation led. (It is important to note that hostiles will use Western "interference" as a rallying point);
5. A smaller force/footprint will be the likeliest most successful approach as opposed to a larger massing of effect.

Needless to say, each of these responses has their own challenges. As such, it is important to examine each point in turn, and lay out the challenges to which it is important to devise solutions in order to address the TCO threat.

Combined, Joint, Interagency

Initially, to be most effective, SOF must operate within a combined, joint, interagency effort that will bring all state power to bear. The threats posed by TCOs cross a number of state responsibilities: financial, legal, environmental, security, and defense. As a result, no one organization has the skillset or, most importantly, the mandate to respond across the full spectrum of all these issues.

Nonetheless, each organization must bring all of their resources to bear for this fight. Imagine a hypothetical situation in Africa in which you have TCOs linked with extremist Islamic organizations laundering money through various legal and quasi-legal business ventures, tied to drugs, counterfeiting, and human smuggling. It becomes immediately evident that a myriad of different government agencies and departments must necessarily be a part of that response.

The solution is not without its challenges, however. After all, it is very difficult to bring the whole-of-government team together. The various departments have different mandates, diverse accountabilities, often different ministers, as well as dissimilar enabling legislation.

Furthermore, it is often not just a simple question of 'stovepipes,' namely a myopic approach to addressing a problem by rejecting collaboration and sharing of information and resources by applying only an 'in-house' approach. There are often excellent and logical reasons specific mandates and restrictions on different organizations exist as they do. For example, privacy and due process are expectations that most people, going about their legitimate business, expect from the state. As such, you cannot simply ignore those demanding policy restrictions.

Another challenge is the planning capacity of most governmental departments. Normally, it is important that the military not be in the driver seat since most of the problems, at their core, are not military problems. However, it must be noted that the employment of military force can be an excellent tool, and the SOF community especially provides sophisticated, targeted, and precise responses to many problem sets.

Nonetheless, despite the reality that it is best not for the military to lead (and be seen as the 800 pound gorilla in the room), too often, by default, it does. This situation is not totally surprising. For example, as commander of the CANSOFCOM, I lead what many military commanders would consider a modest and reasonable planning staff. Yet, compared with the resources available to other government departments, my small staff takes on the image of a planning machine. Considering the experts on my staff who focus on thinking about generating personnel for operations, assembling the necessary intelligence, planning the operation, organizing the logistics to put into place, and projecting out to the next steps, I have many more resources to throw at the planning challenge than my counterpart in the Department of Finance, for instance.

Another challenge in the whole-of-government approach is the creation of additional bureaucracy, namely, creating a combined joint team that simply brings together different organizations that must still report back through their established command chains and require approval for all actions. This process does not necessarily create efficiency or effectiveness. Arguably, it merely creates another level of bureaucracy as each governmental department must work through its own hierarchical structure to achieve permissions.

Once again, as daunting as these challenges are, there are solutions. First, we need to adopt a fully embedded and cooperative approach across the interagency spectrum. This approach demands that the required mandates and authorities are adjusted, amended, and approved by the respective chain of command. In addition, it necessitates the continuation of effort at the operational level. Specifically, mandates and authorities granted from the strategic level must empower the tactical level and give space for operational level planning.

Moreover, it is also important to develop trust between the myriad of organizations that must cooperate. This cooperation demands building and maintaining relationships before they are called upon to act. It also means that we must be careful not to simply replicate existing organizations or create new bureaucracies. Instead, cooperative training and exercises, as well as shared professional development, will promote cooperation and trust. The sharing of standard operating procedures and tactics, techniques, and procedures will "grease the skids" that enable operations. Finally, clear and comprehensive mandates will enable everyone to train, prepare, plan, and work together in an efficient and effective manner.

The Global SOF Network

The next step, or solution, to the array of problems that face us in the complex COE is the utilization of the Global SOF Network, which allows allies to assist each other and friendly states, as well as to build off each other's current engagements in problem areas and in at-risk states. Undeniably, the Global SOF Network concept has evolved from a 'feel-good bumper sticker' to an indispensable force multiplier. The network allows us to apply respective strengths and capacity to address different elements of a specific problem and 'burden share' as needed. This cooperative approach also allows the authorities and abilities of different parts of the various national teams to reinforce and support others who would not normally have direct interactions in these countries or with the myriad of partners.

However, once again, despite the power of the Global SOF Network, we cannot simply ignore national mandates and sovereignty issues. Governments will still be restricted by their own legal limitations. For example, the U.S. Leahy amendment prevents the U.S. military from assisting foreign military units accused of human rights violations. Obviously, no state wants to associate its military assistance with supporting war criminals. However, there is also an understanding that often times those countries most in need of military support tend to have difficulties across the human security spectrum. This becomes a difficult issue to deal with. Do you abandon a state at risk due to past human rights abuse allegations and allow it to slip into greater chaos and humanitarian disaster, or do you work with the government in place and try to institute the necessary reforms?

Canada, for its part, has a number of restrictions on donating lethal aid. Military planners should not look to evade the directions of their governments. Nonetheless, they should understand where our restrictions apply, and seek to address the seams and gaps while respecting sovereign decisions. This solution also speaks to a multi-disciplinary, integrated approach where task-tailored teams attempt to assist with more than just military training.

Regardless of the plethora of issues or location, in many ways a key component to addressing and working at solving many of the problems that exist is trust—both internationally and domestically. Trust is developed by repeated interaction at all levels. Unit commanders need regular meetings with all those they deal with on a regular basis. SOF commanders need to build the trust and expectations with their counterparts, whether in a joint, integrated, or alliance/coalition context. This important bond can only be achieved through regular personal engagements.

There are a number of fora for these persistent engagements. They may include: NATO SOF Headquarters regular commanders' meetings; bilateral and multilateral partner gatherings; and periodic Five Eyes engagements.[3] Within an integrated, whole-of-government context, trust can be built through regular office calls, joint exercises, shared professional development, and working groups. All of these venues work toward building trust and integrated capability.

The Global SOF Network can help to engender the trust which is critical for working effectively with our colleagues in host nations. Those states being assisted must see themselves as included in the solution. Importantly, very few states are willing to admit that they lack the capacity to resolve their own problems, though they may be quick to own up to a lack of resources. Trust can help people work through this issue and allow for substantive dialogue and discussion. However, trust can only be built through persistent presence. Periodic, one-off visits by personnel not aware of host nation culture, issues, and sensibilities are unlikely to be effective in establishing trust.

The issue of trust has another important nexus. Specifically, domestically, home nation publics need to see that their military forces are being used appropriately. Within the context of SOF, this requirement does not mean revealing every aspect of SOF operations to public scrutiny. It does, however, demand that SOF, as part of their professional responsibilities and obligations, are as open and transparent with the public as possible about their activities and achievements (without compromising operational security or methodologies). Accordingly, CANSOFCOM has sought to find the optimum balance between operational security and public disclosure, primarily regarding our international capacity-building programs to ensure Canadians are aware of the contribution the nation is making. Clearly, every disclosure must be tempered by host nation and allied sensibilities.

Organizational Agility and Flexibility

Agility and flexibility are the next 'solution space.' Undeniably, in the face of rapid change that we are all witnessing on the global stage and within the security environment, individuals, organizations, and states must become more adaptive, agile, and flexible. TCOs have clearly proven themselves to be nimble in this regard. Therefore, our responses must be correspondingly responsive and dexterous.

Obviously, this process is easier said than done. Agility and flexibility require a certain type of personality, namely someone who is comfortable in ambiguous scenarios and willing to take risks. Moreover, it requires training programs that step outside of our traditional and conventional methodologies. Perhaps most difficult, it demands that we push responsibility far down the chain of command, and empower personnel at the tactical level across the entire interagency effort to take advantage of evolving situations.

The 'solution space' resides primarily in how we select and train our people. Within the SOF community we rely on strong conventional forces from which to recruit. Without that solid and robust base, we would never meet SOF requirements.

Additionally, we require enhanced professional military education to ensure that we think beyond the limiting scope of 'national defense' needs and develop a culture of 'national security.' This process does not necessarily mean greater formal education (or credentialing). What it does mean is a more wide-reaching education that focuses on how to think (i.e. critical and creative thinking), and developing greater cultural intelligence, tolerance to other ideas, and outlooks. It also means an emphasis at examining cultural and technological aspects of security issues.

Ultimately, focusing on national security means placing an enormous amount of trust in our people. For example, when I deploy a small SOF team somewhere in the world, I like to remind the team commanders that they receive their authority from me, I receive mine from the CDS, the CDS receives his/hers from the Minister of National Defence (MND), and the MND receives his/hers from Parliament. That means that each team commander, whether a sergeant or a captain, is acting only four steps removed from Parliamentary authority. How do we train for that? In essence, we go beyond just thinking about *how to fight* and expand to thinking about *how to think about how to fight*. How do we get our senior leadership to not only accept, but embrace that process? Again, it is a question of trust, professionalism, and experience.

Locally Driven Solutions

The next solution addresses the reality that any and all responses to assist a country must be locally driven and led. As noted, host nations that are recipients of capacity building and advise-and-assist missions become vulnerable to belligerent accusations of 'Western interference,' which can become a rallying cry to increase internal opposition. After all, TCOs and violent extremists may be able to move through porous borders, but ultimately their actions are going to take place on solid ground. Importantly, that geography is still defined by territorial boundaries and borders and, for the most part, publics see themselves as members of that geographical polity. As such, it is important for those publics to see advancements stemming from their government's leadership and initiatives.

As ideal as host nation leadership is, there are several challenges to this proposition. First, any realistic assessment of capacity-building activities demands that we acknowledge that host nation forces may, in fact, be part of the problem. Divided loyalties, corruption, and co-option, have the potential, if we are not careful, to result in us training the 'bad guys.' In addition, we also have to be aware that capacity building, while delivering a professionalized security force in the short term, may result in that same force seeing itself as the only solution in an otherwise corrupt/incompetent state. Significantly, coups or the overthrowing of a government can create new seams and gaps that can be just as easily exploited by TOCs or other adversaries. For example, one would be hard-pressed to describe Mali's 2012 coup as somehow increasing the government's capability to address its national security threats.

Nonetheless, as imposing as the challenge sounds, there are solutions. To start, we need to be in the right places and addressing the right problems. Capacity building sets the baseline and it is obviously a key ingredient. But, it does not stand alone and it must be flexible. Capacity building in Niger, for example, has had to take a backseat to dealing with the specific immediate threats emerging from Boko Haram. Missions may need to be augmented by more risky and operational advise-and-assist operations to deal with serious threats that are beyond the host nation's capacity to deal with on its own. CANSOFCOM is currently combining these efforts in Iraq alongside many of our coalition allies, for example.

Additionally, SOF must use their built-in integral advantages and embrace intelligence-led operations. While threats converge, intelligence also expands and we need to use all aspects of intelligence gathering, which includes tracking emerging patterns through social media intelligence. As we would conduct a military operation in a theater of war, we must apply all possible intelligence resources, techniques, and innovative thinking to specific problem sets to achieve an effective desired outcome.

Sometimes, Smaller is Better

Finally, the last factor to consider in responding to TCOs is that in many cases, smaller force footprints are more likely to be successful than large massing of forces for effect. On these operations, smaller is often better. As such, SOF teams with precise intelligence infusion and host nation cooperation, provide governments with a highly effective option to assisting with the response to global threat convergence. The smaller footprint allows for greater maneuverability, lower risk, lower cost, a more agile and flexible response, and greater acceptance by host nation governments.

Nonetheless, a smaller footprint is not without its challenges. Even acknowledging that SOF's generally smaller task force size and austere footprint allows for a precise and targeted delivery of resources, these resources have their limits. After all, SOF are not limitless. As such, smaller numbers of qualified and experienced SOF operators and an extremely high operational tempo put great strain on SOF force generation. Significantly, it takes years to generate a SOF operator. While there are many advantages to lower signatures in non-declared theaters, these can often be higher-risk missions with all the implications that entails.

Therefore, the SOF community cannot stand alone from their conventional service brothers, but they must be in a position to address their own unique needs. For CANSOFCOM this means owning the responsibility for the five functional domains: force generation, force development, force management, force sustainment and, of course, force employment. SOF's advantage is the tight, flat hierarchy it brings to solving problems.

Nonetheless, you must invest resources to steward the institution and, with limited resources, SOF cannot be everywhere. We must remain constantly engaged with our conventional services, to be prepared to turn missions over as necessary in order to reserve our SOF for those tasks where they are going to deliver the most advantage.

Concluding Remarks

In the end, TCOs represent a significant threat. They are agile, adaptive, and networked, and continually evolve to best take advantage of globalization, particularly within a local and regional context. As such, we must provide a cohesive, collaborative, equally agile response, which is particularly challenging since our adversaries always have the advantage that they can try and break the rules, whereas the 'good guys' must always follow the letter of the law.

Regardless, when required to act, we must always do so swiftly and with flexibility. In addition, the problem set is persistent and widespread; therefore, it is important to steward resources. We cannot be everywhere doing everything. As a result, we need to identify the real threats and rely on the Global SOF Network.

Furthermore, our actions must be in support of the host nation. Leadership must be from behind or beside, not from out front. The host nation must be seen by its public as a credible, sovereign force working to enhance their country. In the end, success will be derived from tight, flat SOF organizations that have credibility and trust from their own governments and military institutions, as well as host nation institutions and populations.

Endnotes

1. I would like to take the opportunity to thank my Policy Advisor, Greg Witol, for his contributions in producing this chapter. Greg's insights have proven invaluable.
2. The World Health Organization states that, "The urban population in 2014 accounted for 54 percent of the total global population, up from 34 percent in 1960, and continues to grow. The urban population growth, in absolute numbers, is concentrated in the less developed regions of the world. It is estimated that by 2017, even in less developed countries, a majority of people will be living in urban areas." Accessed 23 June 2013 at: http://www.who.int/gho/urban_health/situation_trends/urban_population_growth_text/en/.
3. "Five Eyes" refers to an intelligence sharing alliance between Australia, Canada, New Zealand, the United Kingdom, and the United States.

Chapter 5. Transnational Organized Crime: the SOF Nexus

Colonel Earl Vandahl

The contemporary security landscape is filled with threats, none of which is less pressing than TOC, which represents both a domestic, as well as a regional threat. Although largely a law enforcement issue, there is a role for SOF in filling the gaps and enhancing the networks that may have TOC as their principal *raison d'etre*. This viewpoint is steeped in personal experience in command of the CANSOF contribution to an interagency team responsible for domestic response to weapons of mass destruction, as well as a member of the Strategic Joint Staff of the CAF, which provides our CDS with command support. As part of this support to the CDS, the Strategic Joint Staff interacts with the military operational level and the departmental policy level to seek military and political approval for new missions. As such, there is often a nexus with national security issues such as TOC.

It is within the realms of national defense and national security where the TOC/SOF nexus occurs. In fact, there are three areas that warrant further investigation when considering the use of SOF in the fight against TOC:

1. SOF's general attributes and their potential to exploit opportunities to optimize those attributes for this type of fight;
2. SOF capability areas, such as the ability to gain access to inhospitable terrain, surveillance skills, as well as experience in the cyber domain; and
3. Leveraging the Global SOF Network and other interagency networks.

Colonel Earl Vandahl joined the Canadian Special Operations Forces community in 2001, serving in command and staff positions at the Dwyer Hill Training Center and the Force Development Directorate of Canadian Special Operations Forces Command Headquarters in Ottawa. He currently serves in the Strategic Joint Staff as the section head for Middle East and Africa Plans.

The first idea revolves around the concept that there are a number of general SOF attributes that could support the fight against TOC. Specifically, SOF teams are quite small and present little political risk to the contributing nation, as well as the hosting/supporting nation. This, coupled with SOF's very high readiness, translates into speed of mission approval and deployment. As such, SOF can be forward deployed to deal with TOC issues that may be occurring in other countries, but that have a direct impact on domestic and/or regional security. In the new borderless world, shaping and influencing events offshore are often more cost efficient, timely, and effective than waiting for the problems to occur domestically.

As such, in support of these types of international deployments, the complex Status of Forces Agreements for large numbers of troops can be put aside in favor of other approaches to privileges and immunities that are often much easier and more quickly negotiated. Moreover, SOF's orientation toward modest, enduring relationships yields a considerable persistence in their forward deployments, which cultivates the types of partner investments in the relationships that further support the case for improved privileges and immunities.

SOF have also shown themselves to be particularly adept at military and security capacity building.[1] Therefore, governments funding SOF activity recognize that a long-term commitment of SOF to a country yields positive results in the building of the social contract between partner militaries and the governments they serve. In essence, it strengthens governance. This dimension is particularly important in the fight against transnational crime, as weakened political and bureaucratic structures are highly susceptible to corruption and graft.

SOF also provide considerable agility and flexibility to deliver effects that suit the political reality of the region they operate within. The concept of 'by, with, and through' is an integral part of SOF culture and as such relegates SOF to a position in the background, one they are very comfortable in, which minimizes the potential for conflicts stemming from accusations of foreign interference. This can be a major concern for host nations facing opposition criticism of foreign occupation/control. The weak infrastructure, specifically public communication, in at-risk states makes this issue even more problematic. As such, SOF's small footprint also ensures that when operating internationally, the developing countries' bureaucracies and militaries are not overwhelmed by our involvement. Because of their experience in the foibles of the domestic interagency environment, SOF operators can tailor their planning support to provide optimal outcomes that best suit the programs and resources that are available.

SOF's ability, based on their attributes, to contribute to the fight against TOC is also enhanced by their ability to operate within the interagency domain. This ability could be even further enhanced through the vehicle of secondments and exchanges of officials within law enforcement, customs, and national financial regulators, as well as immigration agencies. SOF also have experience with public-private partnership arrangements in the provision of security and other services, and could therefore bring insight into the optimal use of these types of arrangements to provide support in this line of work.[2]

SOF involvement in fighting TOC is not without limitations, and like any government entity there are pressures on budgets. This can be especially true in areas where departments are loath to risk shedding legacy capabilities or adapting what they do have for fear of being caught in a crisis and then suffering the predictable political and public backlash for not being prepared. Moreover, partnerships are increasingly difficult to develop as interagency partners focus on the delivery of their core mandates, sidelining new initiatives. As a result, the inability to work on new capabilities is often not by choice. Rather it is driven by crises that define the volatile nature of the current security environment, distracting governments from the long-game toward outputs and programs that support the domestic audience and political expediencies. This situation produces an allergic reaction to the notion of interagency leadership, a critical element if any meaningful and long-lasting effects are to be achieved in the battlespace of the future.

One final thought on SOF's general attributes is required. In addressing this type of asymmetric problem, SOF will likely offer the use of asymmetric methodologies. The analysis of the problem itself will be complex as an agreed strategy will be necessary. The complexity and dynamism of the threat may dictate several approaches that could involve the targeting of centers of gravity, nodes, links, and lines of communication, as well as a host of other decisive factors. SOF are comfortable with this type of targeting of networks; however, aligning the interagency partners with this way of thinking may be more demanding as cost/benefit analysis is undertaken, as well as the prosecutorial risk for our partners.[3] Clearly, the types of relationships being discussed here are long-term in nature and demand access to a panoply of capability skill-sets within the interagency domain (and potentially beyond).[4]

The second main idea of this chapter is the fact that SOF currently possess capabilities that can contribute to the fight against TOC. Considerations of the issues raised to this point demonstrate that SOF can be organized in this fight as an extension of the domestic arm of law enforcement for out-of-area or rural domestic settings. The SOF ability to operate in challenging terrain is worthy of mention, as it would otherwise be difficult or impossible for other domestic agencies to maintain, amongst other things, surveillance. Specifically, the terrain being referred to is the harshest of land and/or maritime environments, whether domestically (where national laws permit) or internationally, where skills to operate effectively requires advanced training and experience. In addition, these austere environments also require specialized sustainment capabilities.

Furthermore, SOF skills also extend beyond the austere environments to operations within the types of urban landscapes that are common in developing countries that are not familiar to domestic law enforcement. SOF teams have the ability to operate remote systems and maintain a well-established intelligence reach-back capability that can support biometric, electronic, and other means of exploitation to determine pattern of life (for example) as well as support criminal network analysis. Rapid exploitation of sensitive sites is also a SOF capability, where sophisticated portable analytical capabilities can be used to provide forensic quality results in very short timeframes. These types of capabilities contribute to SOF's ability to disrupt or interdict these types of networks.

Importantly, the opportunities for SOF contributions exist beyond the kinetic and reach into cyber. The cyber domain presents emerging opportunities for SOF, such as the exploitation of open-source information and information operations. Growing intelligence opportunities exist within the massive amounts of open-source data in existence, and our adversaries use this information exceptionally well against us, therefore it is time we used it more capably against them.

However, these areas are not without legal and jurisdictional challenges, and there is a recognized need to institutionalize an ability to get out in front of these issues to resolve them, working with new partners and enhancing our networks. TOC's involvement in human trafficking has huge security, economic, and social implications for all of our nations,[5] but trying to make the case for new, more novel approaches to our political and bureaucratic masters is difficult. There must be a clear and unambiguous case for action and a cost-benefit relationship. To get to this point requires that we move our attention further from the 'alligator closest to the boat'—a very daunting challenge when our senior policymakers are consumed day-to-day by crises—to a more distant strategic horizon. As such, there is a need for new forums, new networks, and better thinking to address emerging security threats.

In Canada, we have begun to try to carve some space in busy calendars to look at the more strategic distant threats, or 'alligators a little farther from the boat.' We have termed this the Strategic Orientation Look Ahead (SOLA), or process, which has been underway for approximately a year.[6] Personally directed by our CDS, the SOLA gathers the service chiefs, the CDS's most senior advisors, as well as the department's senior bureaucrats, to examine a topical issue, with the intent of understanding what the implications might be for the immediate (one to three year) planning horizon. There is only enough white space in the diary of senior leaders to conduct four of these sessions per year, as they have a read-in and usually four hours of discussion per topic. The benefit of this investment in time can be significant in shaping thought, as the analysis is often shared with our networks, allies, and interagency partners, as well as our central agencies, and has even gone as high as the head of state.

The final concept to examine is the building of the networks that have been mentioned often in this chapter. There is a significant body of academic study of networks,[7] which is a worthwhile study as SOF rely hugely on their networks for success. The striking issue is that we put more effort into understanding our adversary's networks than we do in understanding our own national and allied networks. In general, networks are most useful in dealing with complex and unpredictable problems, and if a capability cannot be built organically or contracted from the economy, it should be expected that a network type arrangement will be necessary. Networks can be established from the top-down, the bottom-up, and from somewhere in the middle where groups of like-minded individuals may find themselves together coincidentally, and not through deliberate planning.

Networks are hugely popular right now because they are an effective way to ensure communication among disparate members of the net. There has been much done in the last 20 years in the improvement of disaster response networks, for example, often as a result of painful lessons learned.[8] Crises or special events often result in top-down directed, purpose-built networks that also come with the commensurate funding.

Networks are not without their challenges, however, especially accountability and cultural issues. Information sharing and privacy are a challenge, and the outcry at recently enacted anti-terror legislation in Canada shows only one dimension of the problem.[9] But we need networks, especially in the fight against TOC. After all, you need networks to fight networks; but, building networks takes time, patience, and understanding.

In the end, TOC represents an insidious growing threat to nations and regions. Adaptable, agile, and networked, TOC is a complex adversary that requires a coherent, cohesive, networked response. SOF can assist in this fight. SOF possess specific attributes and capabilities that can assist their interagency partners and allies combat TOC. SOF also bring to the game a powerful global network that can apply their collective strength to battling TOC. The key is tying SOF and all the other interagency and regional partners together in a cohesive manner.

Endnotes

1. Canadian capacity-building efforts are funded through the Department of Foreign Affairs, Trade and Development and delivered as a comprehensive (interagency) effort. The program's objectives have consistently been met, and its funding and mandate recently renewed. See http://www.international.gc.ca/crime/ctcbp-parca/mandate-mandat.aspx?lang=eng.
2. The validation and quality assurance of consular security in foreign missions is an area where SOF have experience. Consistently in support of interagency partners and usually at high-risk missions for the development of statements of work, as well as supporting quality assurance once contracts are let. This may also include validating contractor training.
3. Organized crime exploits weak governance and uses graft and corrupt officials to support its objectives. Partnering with countries that operate in this situation poses challenges and evidentiary rules are no exception. Asset freezes and travel bans are preferred where transparency in a partner's justice system is low.
4. One dimension of this is linked to Canada's anti-bribery law, the *Corruption of Foreign Public Officials Act.* On conviction, it precludes private enterprise from bidding on any government work for a number of years. Most nations have this type of legislation on the books, but Canada differs in the scope of activities that are considered offenses as well as the scale of the penalties. The complexity of pursuing criminal networks in this environment demands a more fulsome appreciation of the second and third order consequences of our actions.

5. See the United Nations report at: https://www.unodc.org/documents/human-trafficking/An_Introduction_to_Human_Trafficking_-_Background_Paper.pdf. The profits to organized crime were pegged at $32 Billion USD in 2008, and the expectation is they are growing. The scale of the profits clearly has the potential to undermine security and societal foundations.
6. The SOLA objectives and process were promulgated at UNCLASSIFIED CANFORGEN 093/15 CDS 023/15 201942Z MAY 15 CDS STRATEGIC OUTLOOK FUNCTION. This message is not available on the Internet.
7. There is a useful overview of networks at: http://www.macrothink.org/journal/index.php/jpag/article/viewFile/4870/4397. Issues such as governance, mandates, funding, bottom-up, top-down, etc., are discussed. This article also provides useful insights: Twitchell, D. G., Bodrero, R., Good, M., & Burk, K. (2007). Overcoming challenges to successful interagency collaboration. *Performance Improvement*, 46(3), 8-15.
8. The premise here is that no single level of government can maintain all the necessary resources and capacities to address every need, especially in a large-scale event, thus a network is established. Even where networks do exist, they need to be revisited and refreshed as authorities and accountabilities change. This was a critical shortfall in the Hurricane Katrina relief effort and a second-order consequence of the establishment of the U.S. Department of Homeland Security.
9. For a good summary of the debate on implications for free speech, preventive detention, and the ability of the Canadian Security Intelligence Service to directly intervene—heretofore the domain of law enforcement—see: http://thewalrus.ca/bill-c-51-the-good-the-bad-and-the-truly-ugly/. Although the legislation was amended, its substance remains intact.

Policy and Strategy

Chapter 6. Counter Transnational Organized Crime Strategic Guidance

Dr. Peter McCabe

A goal of national strategic guidance is to guide the direction and actions of a nation's efforts to achieve an advantageous end-state for a particular national interest. National strategic guidance, or national strategies, should address in some form or fashion a nation's national interests. Much has been written about the national interests: for example, Joseph S. Nye defines national interest as "the set of shared priorities regarding relations with the rest of the world."[1] This simplistic definition is useful in many ways but most importantly because it applies to all nations. While the U.S. focuses much of its foreign policy discourse on national interests, it certainly does not have a monopoly on the concept. As it relates to CTOC, the U.S., Canada, and Mexico share the national interests of secure borders, peaceful relations, and free trade. Unfortunately, defining and prioritizing national interests can be problematic. As the Report from The Commission on America's National Interests (2000), posits, "National interests are the foundation of foreign policy … (however,) even among foreign policy elites, there is widespread confusion and little agreement about U.S. interests today."[2] This discord is not a recent phenomenon, as Samuel Huntington argues, "Without a sure sense of national identity, Americans have become unable to define their national interests, and as a result, subnational commercial interests and

---

Dr. Peter McCabe serves as a Resident Senior Fellow with the JSOU Center for Special Operations Studies and Research. Prior to JSOU, he worked as a Strategic Policy Planner at U.S. Central Command. In 2011, Dr. McCabe retired from the U.S. Air Force as a colonel. He received his Ph.D. in Political Science from the University of Florida in 2010, with a focus on International Relations and Comparative Politics.

transnational and non-national ethnic interests have come to dominate foreign policy."[3]

The purpose of this chapter is to provide an analysis of the CTOC strategic guidance documents for the U.S., Canada, and Mexico as they relate to SOF. Some of the key questions that will be addressed are: Do the current national strategy documents for the U.S., Canada, and Mexico adequately address the CTOC issue for SOF? Is the threat adequately addressed in these strategies? How can the existing strategy be improved in the areas of performance and effectiveness? But before these documents are analyzed, a discussion of strategic guidance needs to take place. Putting the word 'strategic' or 'strategy' on the cover of a document is not enough. There are many documents that claim to be strategic but fail to meet even the most rudimentary definition of a strategy. The next three sections will discuss the U.S., Canada, and Mexico strategic guidance, and specifically how they address the CTOC issue. It should come as no surprise that the U.S. strategic documents are quite detailed and apply a whole-of-government approach to the problem. But even with the volume of U.S. documents focused on the CTOC issue, all three nations could improve their strategic guidance by being more proactive and cooperative with each other.

What it Means to be Strategic

There are volumes written about strategy, and this short section will not attempt to summarize it all. What this section will endeavor to do is provide the reader a sense of what it means to be strategic. Most understand the difference between tactical, operational, and strategic—the various levels of thought whether in planning or in warfare. Simplistically, the strategic level is at the top encompassing the whole, whereas operational and tactical address lower levels and just a portion of the whole. So to be strategic is to address time, place, and technology. Strategy covers the diplomatic, information, economic, military, and cultural aspects, or as the preeminent strategy scholar, Dr. Colin Gray, argues, "Just four words express the core of the matter – (Political) Ends, (Strategic) Ways, (Military) Means, and the Assumptions that inform and can well drive action."[4] Dr. Gray also describes strategy as "the bridge that relates military power to political purpose; it is neither military power *per se* nor political purpose."[5] Unfortunately, today's national strategic documents fall short of Dr. Gray's succinct definitions.

For example, the United States NSS provides an overview of the current administration's priorities and focus areas but falls short of being a strategy because it lacks an explanation of the ends, ways, means, and assumptions. A typical critique of the NSS—this one by Walter Russell Mead on the 2010 NSS—describes it as "less a strategy paper than a statement of faith and a wish list."[6]

There are U.S. strategic documents that come closer to being 'strategic,' as those documents focus on a particular topic. While the NSS is trying to coordinate all instruments of state power for all the national security interests, a document like the U.S. Strategy to Combat Transnational Organized Crime (2011) can be more precise and get closer to addressing Dr. Gray's "four words that inform and drive action." However, even this CTOC strategy has its flaws which will be addressed in the next section.

For the purposes of this chapter, the U.S., Canada, and Mexico strategic documents that will be analyzed will not meet the 'strategic' threshold in Dr. Gray's definition. Though lacking the ends-ways-means construct, these documents provide useful strategic policy guidance. However, these official documents, published by their respective governments, are strategic in the sense that they portray their respective nation's priorities and focus areas. Any mention of combating TOC will display that nation's concern for the issue and any mention of possible solutions will start to address ends, ways, and means. A strategy should translate focus areas into concrete initiatives, but it will become apparent in the following sections that national strategic guidance often falls short.

What does it mean to be strategic? For some, it is theoretical (definitional), while for others it is practical (policies). For our purposes, strategic means providing enough guidance for practitioners at the national level to discern a way ahead. As we will see in the following sections, this low bar for strategic guidance is necessary to avoid getting bogged down in theoretical arguments. Too often critics describe 'how' a document should have been written instead of focusing on 'what' is inside the document. To that end, the next section will address the U.S. CTOC strategic guidance—specifically the 2015 NSS and 2011 Strategy to Combat Transnational Organized Crime.

CTOC Strategic Guidance (United States)

Any discussion of U.S. strategic guidance must start with the fact that there is a proliferation of U.S. national strategic documents. A simple search on the Homeland Security Digital Library comes up with a list of 125 strategic documents[7] covering various topics such as borders and immigration, infrastructure, intelligence, international, law enforcement, maritime, military, national security, pandemics, public health, space, technology, terrorism threats, transportation, weapons, and other areas. A vast bureaucracy supports the United States Government so the enormous supporting documentation should come as no surprise. At the top of those documents is the U.S. NSS.

The NSS is a report required by Federal Law 50 USC 3043 (previously section 404a)[8] which shapes exactly what this 'strategy' must lay out. It is prepared by the executive branch for Congress. In 1986 a law was passed requiring the President to give Congress an annual 'Strategic Statement.' All Presidents are inconsistent in meeting the congressional guidance; for example, George W. Bush only produced two in eight years. President Obama has now produced two while in office (2010 and 2015). The NSS lays out the administration's plans to address national security concerns using all the elements of national power. The document identifies four enduring national interests: security of the U.S., its citizens, U.S. allies and partners; a strong, innovative and growing U.S. economy in an open international economic system that promotes opportunity and prosperity; respect for universal values at home and around the world; and a rules-based international order advanced by U.S. leadership promoting peace, security, and opportunity through stronger cooperation to meet global challenges.[9]

Richard Betts notes the NSS "has sometimes been a Christmas tree on which every interest group hangs its foreign policy concerns,"[10] hence, the lack of 'strategy' in the NSS document. The NSS mentions CTOC five times, which for the top U.S. strategic document is significant. In the introduction, CTOC is mentioned as one of the top strategic risks to U.S. interests. In section two, addressing security, CTOC is mentioned twice. In both instances it refers to combating organized crime in weak and failing states through risk-based approaches. The final two mentions can be found in the international order section, and they address the organized crime threat to U.S. collaborative efforts in the Western Hemisphere. The U.S. takes the TOC threat seriously and the NSS outlines those concerns; however, the NSS falls short of outlining the CTOC ends, ways, and means, and makes no mention of how the U.S. military, let alone SOF, are to be involved. This is not surprising in a 35 page document that provides a vision, outlines interests, and offers vague direction. The next document, the 2011 U.S. Strategy to Combat Transnational Organized Crime, gets closer to being 'strategic' and provides some ends, ways, and means.

The 2011 CTOC strategy is more specific—it focuses solely on one threat whereas the 2015 NSS addresses all threats. The 2011 document is clear: TOC threatens U.S. and international security. TOC threatens rule of law, economic interests, and intellectual property while providing support (facilitators) to terrorist groups, smuggling networks, and cybercrime. To address these threats, the CTOC strategy has five objectives, 56 priority actions, and introduces capabilities and tools. The implementation of this strategy will be an interagency effort.[11]

The Interagency Policy Committee on Illicit Drugs and Transnational Criminal Threats, led by the National Security Staff and the Office of National Drug Control Policy, oversees implementation of the strategy.[12] Some of the more powerful interagency implementation tools include:

- White House Executive Order 13581 (Blocking Property of Transnational Criminal Organizations);
- Department of Treasury Office of Foreign Assets Control Transnational Criminal Organizations Sanctions Program;
- Code of Federal Regulations 590, Transnational Criminal Organizations Sanctions Regulations;
- Presidential Proclamation (July 25, 2011): Suspension of Entry of Aliens Subject to United Nations Security Council Travel Bans, and International Emergency Economic Powers Act Sanctions; and
- Department of State, Transnational Organized Crime Rewards Program.

So how do SOF fit into the 2011 CTOC strategy? Or does it only apply to the Departments of State, Justice, and Treasury? On the contrary, SOF have a role in this strategy—albeit a supporting role to law enforcement. The interagency Threat Mitigation Working Group is one example. This National Security Council-sponsored group produces action plans for high-value targets and has Office of Secretary of Defense and Joint Staff representatives. The military and specifically SOF have a strong role in technical assistance and capacity building. While the Department of Justice has the lead to work law enforcement with foreign partners, DOD, using SOF, assists with building partner capacity (BPC). The shortfall in these efforts resides in the fact that no real operational entity exists to prioritize and synchronize U.S. CTOC efforts.[13] This issue will be addressed later in the section.

The SOF role in CTOC is small but growing. Currently, SOF build partner capacity in many states that are engaging in the CTOC fight. As Dr. Harry Yarger argues in his recent JSOU monograph, "Under the conditions of rapid and continued globalization, the lynchpin of the emerging United States grand strategy is building partner capacity (BPC), and SOF are instrumental in the pursuit of a successful BPC policy."[14] SOF participate in BPC through the FID mission, which is "the participation by civilian and military agencies of a government in any of the action programs taken by another government or other designated organization to free and protect its society from subversion, lawlessness, insurgency, terrorism, and other threats to their security."[15] SOF assisting the security forces of fragile or weak states is a piece (albeit small) of the CTOC puzzle.

Another piece of the puzzle is the nexus of crime and terrorism as addressed by the CTOC strategy: "The Department of Justice reports that 29 of the 63 organizations on its FY2010 Consolidated Priority Organization Targets list … were associated with terrorist groups."[16] This convergence of crime and terrorism is a debatable concept, but even if the link is tenuous at best, the SOF role in counterterrorism still has an impact in the CTOC strategy. Mexico is one example used in the CTOC strategy: "TOC in Mexico makes the U.S. border more vulnerable because it creates and maintains illicit corridors for border crossings that can be employed by other secondary criminal or terrorist actors or organizations."[17] To the end, "much of U.S. military and security assistance has been targeted at the violence and trafficking of some seven major drug trafficking organizations as well as smaller operations."[18] While the CTOC strategy does not explicitly address SOF participation, it does identify one of its priorities as, "enhancing Department of Defense support to U.S. law enforcement through the Narcotics and Transnational Crime Support Center."[19] Therefore, the U.S. national security strategies (NSS and CTOC) adequately address the CTOC issues and threats, but SOF must interpret their CTOC role vice draw from specific strategic guidance. As previously noted, the NSS provides a cursory overview of the organized crime threat, mainly as a strategic risk to national interests. TOC has a significant security impact on weak and failing states, which disrupts the U.S. economy and foreign policy. The CTOC strategy is more specific in outlining the threat. The strategy outlines how organized crime penetrates state institutions, encourages corruption, threatens governance, and damages the U.S. economy, competitiveness, and strategic markets. Organized crime accomplishes this through trafficking (drugs, humans, weapons, intellectual property), cybercrime, and facilitation of those who would carry out violent acts (terrorists and other criminals). While both strategies address the CTOC issues and threats, there are areas of improvement for both in measures of performance and effectiveness.

The main critique of the NSS revolves around the lack of specificity within the document. Every administration faces the challenge of creating a coherent strategy and maintaining a consensus about how to protect national interests. The solution that each administration employs is to create a document that covers enough issues but reduces the specificity to achieve the lowest common denominator. As Richard Doyle notes, what is eventually produced is "an NSS that offends none of the important participants by saying little of significance."[20] The NSS can be improved by providing measures of performance and effectiveness. Since there have been only four NSSs produced in the last 13 years (2015, 2010, 2006, 2002), enough time lapses between each NSS to conduct an assessment. These measures could assess whether the top strategic risks increase, decrease, or remain the same. The NSS could include measures on resource allocation. The CTOC efforts already mentioned are interagency solutions and require coordinated efforts among disparate organizations. Does law enforcement have enough resources to lead the CTOC effort? These measures of performance and effectiveness can be applied to the CTOC strategy.

The CTOC strategy already contains objectives, actions, capabilities, and tools. While more specific and closer to being a strategy than the NSS, the CTOC strategy could also be improved with measures of effectiveness and performance. In the priority actions section, providing more specificity to the identified actions could make them measurable. For example, the first action mentioned for the "Start at Home" objective is, "Reduce the demand for illicit drugs in the United States, thereby denying funding for illicit trafficking organizations."[21] This action could be measurable by changing it to read: reduce the demand for illicit drugs in the United States by 50 percent by 2020, thereby denying funding for illicit trafficking organizations. Each of the actions within the strategy can be modified in this way to add the necessary specificity and assessment aspect to the document. As John Driscoll argues, "proper metrics to measure progress ... will need to be employed to ensure accurate measurement of ultimate goals, not merely intermediate objectives."[22] Both the NSS and CTOC strategies require specificity added to the existing vision and direction. This argument also applies to other states as well. The next section will address Canada's CTOC strategic guidance.

CTOC Strategic Guidance (Canada)

Much like the United States, Canada's fight against TOC is problematic, not only because of the number of illicit activities involved, but also because of the number of agencies (local, provincial, and federal levels) assigned to address the threat. Gavril Paraschiv posits, "Transnational organized crime is a phenomenon that has emerged in different cultures and countries around the world: it is a new category of crimes, being a significant global threat."[23] The countries in the Western Hemisphere and specifically North America are not exempt from this threat. In fact, with the long borders between Canada and the U.S. (longest border between two countries in the world—5,525 miles), and Mexico and the U.S. (1,933 miles), they make combating TOC a momentous challenge.[24]

One of Canada's overarching defense strategies is a legacy document (published in 2008) titled, *Canada First Defence Strategy* (CFDS). Much like the United States' NSS, the CFDS uses a whole-of-government approach to meet its domestic and international security requirements. The CFDS outlines three roles for Canadian Forces (defending Canada, North America, and abroad); six core missions (all focused on safekeeping the populace), and four pillars (personnel, equipment, readiness, and infrastructure). This strategic document is much more useful than the NSS due to its detailed modernization and investment sections (programmatic funding details to 2028). However, its treatment of the strategic environment is similar to the NSS in its broad outline of potential threats. The only mention of organized crime is as follows: "Ethnic and border conflicts, fragile states, resurgent nationalism and global criminal networks continue to threaten international stability."[25] This single entry in the CFDS and lack of strategic direction on how it is to be addressed does not provide SOF the necessary guidance it requires.

One of the main obstacles that Canadian SOF encounter is the jurisdictional issue that creates a barrier to any integrated CTOC approach. Much like the U.S. *Posse Comitatus* law (limit the powers of the federal government in using its military personnel to act as domestic law enforcement personnel), Canada's history and constitution limit the federal powers. The provinces retain certain powers in "areas that pertain to national security yet the criminal code constitutionally comes under federal jurisdiction, and the Royal Canadian Mounted Police (RCMP) is a federal police force."[26] Therefore the CTOC effort, specifically the SOF role, must tread carefully to avoid perceptions of interference and/or crossing jurisdictional lines. To aid in the effort of interagency cooperation,

> Regional Integrated National Security Enforcement Teams (INSETs) have been established across the various regions of the country to bring together the RCMP, provincial police services (in Ontario and Quebec), municipal police departments, and the Canadian Security Intelligence Service (CSIS) in order to ensure that operational investigations relating to transnational crime and terrorism are properly coordinated.[27]

Note this does not include the Canadian military (SOF) which would only be drawn in on an as needed basis.

Similar to the U.S., Canada sees the CTOC effort in terms of the crime-terror nexus. In Canada, section 83.01 of the Criminal Code defines terrorism as an act committed "in whole or in part for a political, religious or ideological purpose, objective or cause ..." with the intention of intimidating the public "... with regard to its security, including its economic security, or compelling a person, a government or a domestic or an international organization to do or to refrain from doing any act."[28] The definition of an organized crime group, as stated in the Criminal Code, is a group which is composed of three or more persons in or outside Canada, and has as one of its main purposes or main activities the facilitation or commission of one or more serious offenses, that, if committed, would likely result in direct or indirect receipt of a material benefit, including a financial benefit, by the group or by any one of the persons who constitute the group.[29]

There are numerous documents and reports that address CTOC and the crime-terror nexus, but they are all law enforcement focused. One example is the Public Report on Actions under the National Agenda to Combat Organized Crime, *Working Together to Combat Organized Crime*, which overviews the scope of serious and organized crime in Canada, highlights governments' collective response to this problem, discusses where action should focus, and identifies strategies and approaches recommended by the National Coordinating Committee on Organized Crime to reduce its harms.[30] Another is the *Canadian Law Enforcement Strategy on Organized Crime*, which reflects a national collaborative effort of intelligence and operations to detect, reduce, and prevent organized and serious crime.[31] *Canada's Strategy for Engagement in the Americas* is another whole-of-government approach with three broad goals: first, increasing Canadian and hemispheric economic opportunity; second, addressing insecurity and advancing freedom, democracy, human rights, and the rule of law; and third, fostering lasting relationships.[32] Each of these documents provides guidance to policymakers but is focused on law enforcement agencies exclusively.

Despite the lack of Canadian CTOC strategic guidance for SOF, there are opportunities to explore. Nicholas Dorn argues the operating space for CTOC can encompass all instruments of national power. This operating space includes law enforcement (policing leading to judicial decision making); administrative measures (confiscation, fines, and civic sanctions); market regulation (suspension of commercial rights/privileges); and disruption tactics (includes police, customs, and other security agencies).[33] This last category is the SOF operating space. Again, due to Canada's constitutional restrictions, SOF must tread lightly and cannot act unilaterally. Working relationships between government agencies are based on trust that needs to develop over time. While law enforcement can and do conduct disruption tactics, SOF can cooperate and participate in this CTOC role. The Canadian military (SOF) know all too well the price that is paid for lack of attention on the crime-terror nexus. On 20 and 22 October 2014, two Canadian Forces soldiers were murdered in separate terrorist attacks that took place outside Montreal and in downtown Ottawa, respectively. While both of these incidents were conducted by individuals who had become radicalized and motivated by violent extremist ideology, it is a stark reminder that Canadian law enforcement and homeland security entities (military) need to work together.

Specifically, CANSOFCOM's primary mission is counterterrorism which involves working with law enforcement agencies to protect the populace.

A different perspective on the threat to Canada and its populace comes from Major Bernard Brister (Royal Military College of Canada) who argues that Canadian security interests should focus on Mexico. Major Brister contends that internal Mexican affairs influence Canadian security in three ways: immigration, illegal drugs, and security relationships. On immigration, "convinced that the domestic economic situation within Mexico is irreversible and/or fearing for their lives as a result of the increasing dominance of criminal organizations in many parts of everyday Mexican life, significant numbers of Mexican citizens began making their way to Canadian ports of entry and claiming refugee status."[34] While Brister acknowledges that the flow of drugs into Canada from Mexico is just a fraction of what travels to the U.S., the "volume of cocaine smuggled into Canada has tripled in recent years."[35] Finally, Brister argues the greatest threat is the effect these issues are having on the U.S.-Canada security relationship. Brister uses the term "thickening of the border" in reference to U.S. efforts to "fire-proof their citizens from the triple threat of illegal immigration, the importation of increasing amounts of illegal narcotics, and terrorist attack."[36] These concerns are valid as TOC takes advantage by moving people, drugs, and guns across borders and can be mitigated through international and interagency coordination that involves the military (SOF).

While the U.S. and Canada have a strong working relationship in their CTOC efforts, the next section will highlight the Mexican strategic document deemphasizing cross border interactions.

CTOC Strategic Guidance (Mexico)

Similar to the U.S. and Canada, Mexico's fight against organized crime is complicated by a variety of factors (geography, limited resources, and external demand for drugs). Geography plays a major role with land borders north (with the U.S. 3,155 km) and south (Belize 276 km and Guatemala 958 km); ocean coastlines east and west (total of 9,330 km) with terrain that varies from rugged mountains to low coastal plains, high plateaus, and desert.[37] This unique geography challenges Mexico's security apparatus to secure the borders from smuggling and other nefarious activities. Another factor is the limited resources Mexico possesses in comparison to the U.S. and Canada.

While Mexico has a $1.3 trillion economy, per capita income is roughly one-third that of the United States and two-fifths that of Canada. Mexico's economy "is vulnerable to global economic pressures, such as lower external demand, rising interest rates, and low oil prices—approximately 30 percent of government revenue comes from the state-owned oil company, PEMEX."[38] This disparity in resources is even more evident in military expenditures. Using 2012 data, military spending as a percentage of GDP is as follows: U.S. 4.35, Canada 1.24, and Mexico 0.59.[39]

Mexico has almost no control over the external demand for drugs and other smuggled items/people. The demand for these illicit items is growing, both in the U.S. and Canada. As a result, TCOs continue to battle with each other and the Mexican government security forces. "Following a decision by the former administration to directly confront [TCOs], there have been 50,000 or more narco-related homicides since 2007."[40] To address this level of violence and outline a plan of action, the current Mexican administration produced a strategy document titled, "National Security Program (NSP) 2014-2018: A multidimensional policy for Mexico in the 21st century."

The NSP is a very detailed and comprehensive document especially compared to the U.S. NSS and Canadian CFDS. It contains four sections: the administration's policy outline, strategic environment, strategic objectives (goals and lines of action), and long-term trends. The policy outline includes the legal and conceptual framework that references the Mexican Constitution. The overarching vision (Mexico in peace) promotes a national security policy that guarantees internal and external security. This is achieved through the National Security Council (NSC) which consists of the Ministry of the Interior, Secretary of National Defense, Secretary of the Navy, Secretary of Public Administration, Ministry of Foreign Affairs, Secretary of Communications and Transport, and Attorney General. The approach to national security is to "promote the security of Mexico through a multidimensional policy that anticipates those internal and external trends that can jeopardize [the] nation … thus safeguarding freedom, human rights and the security of … citizens."[41] The goal of the NSC is to integrate these institutions to promote cooperation and a culture of safety.

The Mexico strategic environment is outlined in the next section of the NSP. It argues for a position in the world that promotes the advances of Mexico's economy but recognizes the enormous challenges the country faces (internal security). The NSP highlights Mexico's role in North America and its partnership with the U.S. and Canada. There are numerous bilateral and trilateral agreements between the nations, but the North American Free Trade Agreement is the most famous. According to the latest data, between 1993 and 2012, trade among the three members quadrupled from $297 billion to $1.6 trillion.[42] The NSP highlights the importance of the North American relationship and the efforts for institutional cooperation and exchange. That is certainly true for economic, administrative, health, and other domestic issues; however, in the realm of security and specifically in CTOC efforts, the trilateral efforts have been wanting. The role of the Mexican armed forces in the maintenance of internal security has been significant. While criminal groups still have economic power, armament, and territorial presence, the armed forces (including SOF) will continue to carry out the following types of operations: reduce violence; cultivation eradication; inhibit trafficking (weapons, drugs, people); and increase strategic installation safety. These operations could be enhanced with increased cooperation among American, Canadian, and Mexican SOF.

When addressing risks and threats, the NSP contains a section dedicated to TCO. The NSP acknowledges that domestically the actions of organized crime has had a corrosive effect on the confidence of the society in the institutions, the maintenance of democratic governance in specific regions of the country, and the economic and social development in those regions.[43] Internationally, the highly publicized increase in violence in the country has negatively impacted the image of Mexico abroad. Organized crime groups persist in maintaining their presence in specific regions of the country conducting their illicit activities and using violence to ensure the continuity of their activities and transfer and sale of drugs.

Based on the Mexican CTOC efforts, the NSP outlines the following trends: change in land and air routes of drug trafficking; the development of the internal market for the consumption of illegal drugs in the country; the change in consumption patterns; the diversification of the criminal activities of the criminal groups; and the expansion of their sources of income.[44] While Mexican CTOC efforts have had a fragmentation effect on organized crime, it has fallen short of arresting the violence and reducing the threat.

The NSP seeks to codify federal actions (State and municipal) to confront the situation from a regional perspective by including local governments in the effort. The key to success is fusing intelligence from various sources that can provide early warning of illicit activities. Specifically against drug traffickers, the NSP argues for a whole-of-government response that includes the Secretary of the Interior, Secretary of National Defense, Secretary of the Navy, and the Office of Attorney General. Unlike the U.S. and Canada, Mexican SOF have a central role in CTOC efforts. What is missing from the NSP strategy is an international coalition approach to CTOC.

The third section of the NSP outlines the goals, strategies, and lines of action required to achieve a secure internal and external security. Strategic objective 2 states, "Make sure that the Mexican State national security policy adopts a multidimensional perspective through the coordination of authorities and institutions … to favor the achievement of the objectives and national interests."[45] Clearly, this objective emphasizes cooperation among internal security organizations. A sub-objective could include an external component that underscores the external aspect among the U.S., Canada, and Mexico. Strategic objective 2.2, "Strengthening the response capacity of the Federal forces to contribute [to] the maintenance of internal security as well as the tasks of exterior defense of the Federation,"[46] gets closer to the goal of cooperative coalition actions. This objective can be achieved through military education and training programs, combined military exercises, and exchange programs. Trilateral SOF cooperation could assist in addressing issues of TOC affecting the three neighboring countries through robust intelligence sharing, BPC, and security cooperation activities.

The NSP's fourth section concludes by looking to the future and the national security challenges facing Mexico. The focus of this section is on long-term issues that could impede sustainable development of Mexican society. The four challenges include: the preservation of biodiversity as a strategic resource; impacts of climate change on food security and the management of water; transformation of the global energy outlook and energy security; and management of the health risks of a global pandemic.[47] While the NSP does not address organized crime in these four challenges, it is not difficult to appreciate how organized crime can take advantage of each situation. Developing solutions to Mexico's challenges requires a collaborative approach requiring innovation. No one agency can accomplish the task, and only working in concert with other nations (U.S. and Canada) can Mexico hope to address these future challenges. SOF have a role today in combating TOC and a role tomorrow to develop sustainable solutions.

Conclusion

The CTOC authoritative, legal, and operational challenges facing the U.S., Canada, and Mexico are daunting. As Naim observed in 2005, "more than 90,000 merchant and passenger ships dock at U.S. ports. They carry 9 million containers with 400 million tons of cargo. Another 157,000 smaller vessels call at U.S. harbors. The notion that a government agency ... can seal such a porous border in this era is challenging, to say the least."[48] These same facts and figures can be applied to Canadian and Mexican ports of entry and border crossings (official and unofficial). SOF have a CTOC role—albeit there are legal restrictions on how much U.S. and Canadian SOF can get involved. The U.S., Canada, and Mexico strategic documents reflect those restrictions. The U.S. NSS mentions CTOC but does not mention SOF, while the 2011 CTOC strategy is law enforcement focused and hence the military (SOF) have a small supporting role only. The USSOF role resides in BPC and in the crime-terror nexus. Canada's CFDS only mentions organized crime once, and while a more comprehensive strategy than the United States' NSS, it still falls short of providing SOF the necessary guidance. Canadian strategy documents focus on the terror-crime nexus, but these documents are focused on law enforcement and not SOF. The constitutional restrictions on Canadian SOF limit their participation in CTOC; CANSOFCOM can focus on CT (a primary mission) and work with law enforcement, as well as U.S. and Mexican SOF. Finally, Mexican SOF have the most experience in CTOC efforts. The U.S. and Canadian SOF can learn much from their counterparts to the south. In addition, the well-trained U.S. and Canadian SOF can assist their Mexican counterparts in achieving the NSP goals, strategies, and objectives. Stability and security in North America is an overarching goal for all three nations and is identified as such in the respective strategic documents. Only through seeking opportunities to work together on a variety of issues, not just CTOC, will this be achieved.

Endnotes

1. Joseph S. Nye Jr, "Redefining the National Interest," *Foreign Affairs* (Vol. 78, No. 4, Jul. - Aug., 1999): 23.
2. Commission on America's National Interests (July 2000), 11, accessed 3 June 2015 at: http://belfercenter.ksg.harvard.edu/files/amernatinter.pdf.
3. Samuel Huntington, "The Erosion of American National Interests," *Foreign Affairs* (Vol. 76, No. 5, Sep. - Oct., 1997), 29.

4. Colin S. Gray, "Strategy and Security," *Infinity Journal* (Volume 4, Issue 3, Spring 2015), 13.
5. Colin S. Gray, *Modern Strategy* (Oxford: Oxford University Press, 1999), 17.
6. Walter Russell Mead, "Review of the National Security Strategy," *Foreign Affairs* (Vol. 89, No. 5, September/October 2010), 158-159.
7. Homeland Security Digital Library, National Security Documents, accessed 30 May 2015 at: https://www.hsdl.org/?collection&id=4.
8. Detailed requirements of the National Security Strategy can be found at the U.S. House of Representatives website, Annual National Security Strategy Report, accessed 5 June 2015 at: http://uscode.house.gov/view.xhtml?req=granuleid:USC-prelim-title50-section3043&num=0&edition=prelim.
9. National Security Strategy Archive, accessed 5 June 2015 at: http://nssarchive.us/.
10. Richard K. Betts, "U.S. National Security Strategy: Lenses and Landmarks," November 2004, accessed 4 June 2015 at: http://www.princeton.edu/~ppns/papers/betts.pdf.
11. Barack Obama, *Strategy to Combat Transnational Organized Crime* (July 2011), National Security Council website, accessed 10 May 2015 at: https://www.whitehouse.gov/sites/default/files/microsites/2011-strategy-combat-transnational-organized-crime.pdf.
12. The White House, "Fact Sheet: Strategy to Combat Transnational Organized Crime (July 25, 2011)," accessed 10 May 2015 at: http://iipdigital.usembassy.gov/st/english/texttrans/2011/07/20110725111924su0.6405407.html.
13. Interview with Special Forces Colonel Clyde A. Moore, USSOCOM Senior Liaison Officer, U.S. Department of Justice, 20 May 2015.
14. Harry R. Yarger, "Building Partner Capacity," *JSOU Press* (Report 15-1, February 2015), 1.
15. Joint Publication (JP) 3-22, Foreign Internal Defense, ix.
16. Obama, Strategy to Combat Transnational Organized Crime, 6.
17. Ibid.
18. Graham H. Turbiville, Jr., "U.S. Military Engagement with Mexico: Uneasy Past and Challenging Future," *JSOU Press* (Report 10-2, March 2010), 45.
19. Obama, *Strategy to Combat Transnational Organized Crime*, 18.
20. Richard B. Doyle, "The U.S. National Security Strategy: Policy, Process, Problems," *Public Administration Review* (July/August 2007), 628.
21. Obama, *Strategy to Combat Transnational Organized Crime*, 16.
22. John Driscoll, "Improving U.S. Strategy on Transnational Organized Crime," *International Affairs Review* (Vol. 23, Number 1, Fall 2014), 100.
23. Gavril Paraschiv, "Conceptualizing Transnational Organized Crime," *Economics, Management and Financial Markets* (Vol. 8 (2), 2013), 175.

24. U.S. Census Bureau, *Statistical Abstract of the United States: 2012*, accessed 24 June 2015 at: http://www.census.gov/compendia/statab/2012/tables/12s0363.pdf.
25. Government of Canada, *Canada First Defense Strategy*, 6, accessed 28 June 2015 at: http://www.forces.gc.ca/assets/FORCES_Internet/docs/en/about/CFDS-SDCD-eng.pdf.
26. Felia Allum and Stan Gilmour, *Routledge Handbook of Transnational Organized Crime* (New York: Routledge, 2012), 414.
27. Ibid., 415.
28. Government of Canada, *The Criminal Code of Canada*, R.S.C. 1985, c. C-46, s. 83.01. Current to 9 June 2015.
29. Ibid.
30. Government of Canada, *Working Together To Combat Organized Crime*, 1, accessed 28 June 2015 at: http://www.publicsafety.gc.ca/cnt/rsrcs/pblctns/cmbtng-rgnzd-crm/cmbtng-rgnzd-crm-eng.pdf.
31. Government of Canada, Criminal Intelligence Service Canada website, accessed 28 June 2015 at: http://www.cisc.gc.ca/media/2014/2014-08-22-eng.htm.
32. Government of Canada, *Canada's Strategy for Engagement in the Americas*, accessed 28 June 2015 at: www.americas.gc.ca.
33. Adam Edwards and Peter Gill, *Transnational Organized Crime: Perspectives On Global Security* (New York: Routledge, 2003), 230.
34. Brister, Bernard, *Forget Al-Qaeda: Think Mexico - The Next Great Threat to Canadian Security*, (Calgary: Canadian Defence & Foreign Affairs Institute, 2012), 2-3.
35. Ibid., 3.
36. Ibid., 4.
37. The World Factbook, CIA website, accessed 30 June 2015 at: https://www.cia.gov/library/publications/the-world-factbook/geos/mx.html.
38. Ibid.
39. Ibid.
40. Integrated Country Strategy, Mexico FY2014-2016, 7, accessed 30 June 2015 at: https://intellipedia.intelink.gov/w/images/4/41/Mexico_Integrated_Country_Strategy1_2014-16.pdf.
41. Mexico Government, Secretary of the Interior (SEGOB), Official Journal of the Federation, accessed 23 March 2015 at: http://dof.gob.mx/nota_detalle.php?codigo=5342824&fecha=30/04/2014.
42. Office of the U.S. Trade Representative, NAFTA, accessed 30 June 2015 at: https://ustr.gov/trade-agreements/free-trade-agreements/north-american-free-trade-agreement-nafta.
43. Mexico Government, Secretary of the Interior (SEGOB), Official Journal of the Federation, accessed 23 March 2015 at: http://dof.gob.mx/nota_detalle.php?codigo=5342824&fecha=30/04/2014.

44. Ibid.
45. Ibid.
46. Ibid.
47. Ibid.
48. Moises Naim, *Illicit: How Smugglers, Traffickers, and Copycats are Hijacking the Global Economy* (New York, NY: Anchor Books, 2005).

Chapter 7. North American Efforts to Combat the Financing of Terrorism

Professor Celina Realuyo[1]

Counterterrorism has become once again a primary national security concern for the U.S. and its partner nations. Last year (2014) marked the deadliest year from terrorist attacks since such statistics have been compiled in the early 1970s,[2] with the brutal acts of violence perpetrated by terrorist groups like Islamic State of Iraq and the Levant (ISIL), Boko Haram, and al-Shabaab dominating the headlines. North America was not immune to terrorism with the October 2014 terrorist attacks in Canada, scores of arrests in the U.S. and Canada that foiled potential terror plots, and evidence of North American citizens traveling to Iraq and Syria to join ISIL as foreign fighters. Despite major national security efforts to address the terrorist threat, including the creation of the U.S. Department of Homeland Security (DHS) and U.S. Northern Command, governments are still struggling to better understand, undermine, and counter the scourge of terrorism in this post-9/11 global security environment.

Terrorist groups are motivated by ideology and political aspirations, in contrast to criminal organizations that are driven by greed. Terrorists and criminals both require multiple enablers to support their networks and realize their agendas. These critical enablers include leadership, personnel, weapons, logistics, illicit activities, corruption, and financing. Of all these, perhaps the most vital is financing, as all these critical enablers require funding. Consequently, the financial front to combat terrorism is a crucial

---

Celina Realuyo is a JSOU Senior Fellow and is Professor of Practice at the William J. Perry Center for Hemispheric Defense Studies at the National Defense University where she focuses on U.S. national security, illicit networks, transnational organized crime, counterterrorism, and threat finance issues in the Americas. She is a former U.S. diplomat, banker with Goldman Sachs International, U.S. State Department Director of Counterterrorism Finance Programs, and professor of international security affairs at the National Defense, Georgetown, and George Washington Universities.

component of CT campaigns, such as the current endeavor to degrade and defeat the Islamic State.

To sustain themselves and conduct their operations, terrorist groups need to raise, move, store, and spend money. Understanding how these groups are financed is instrumental in devising strategies to counter and neutralize them. In May 2015, as part of Operation Inherent Resolve to counter ISIL, USSOF conducted a daring raid in Syria against Abu Sayyaf, a senior leader considered the chief financial officer of ISIL.[3] ISIL is considered the richest terrorist group in history, and this operation illustrates the growing importance of targeting the group's finances and how valuable the financial intelligence collected at the target site could be to attack ISIL's networks.

Since the tragic attacks of 11 September 2001, Canada, Mexico, and the U.S. have developed the necessary legal frameworks and effective mechanisms to detect, disrupt, and deter the financing of terrorism but must remain vigilant and keep in step with emerging financial technologies that could be used by terrorists and criminals to fund their nefarious activities. This chapter will examine the series of legal, law enforcement, military, and intelligence measures instituted to counter the financing of terrorism in Canada, Mexico, and the United States. It concludes with a description of the global campaign to degrade and defeat ISIL to illustrate the importance of the financial instrument of national power in current CT efforts at home and abroad.

Canada

Canada has been proactive in detecting and preventing terror plots in the post-9/11 era, but the terrorist attacks the country suffered in 2014 and disturbing number of Canadian ISIL sympathizers and foreign fighter recruits have dramatically changed its CT posture. On 20 October 2014, Martin Couture-Rouleau, a 25-year-old Québécois recent Muslim convert and ISIL supporter, used his car to run down two soldiers in front of a federal building in Saint-Jean-sur-Richelieu before being fatally shot by police after an ensuing car chase; one of the soldiers, Warrant Officer Patrice Vincent, subsequently died from his injuries. Two days later, on 22 October 2014, a lone gunman inspired by Islamic extremism attacked Parliament Hill, leaving one honor guard soldier dead and Canada in shock. Although it does not appear the two attacks were related, they served as a dramatic wake-up call that Canada was not immune to Islamic terrorism. Prime Minister Stephen Harper called the shootings a "terrorist act" and stated:

> This will lead us to strengthen our resolve and redouble our efforts and those of our national security agencies to take all necessary steps to identify and counter threats and keep Canada safe here at home, just as it will lead us to strengthen our resolve and redouble our efforts to work with our allies around the world and fight against the terrorist organizations who brutalize those in other countries with a hope.[4]

In response to these terrorist attacks, Canada is significantly enhancing measures to counter terrorism and strengthen its national security with new legislation known as Bill C-51, the Anti-Terrorism Act of 2015: Protecting Canadians from National Security Threats at Home and Abroad. This is the first comprehensive security reform of its kind since 2001. The controversial legislation would expand the powers of Canada's spy agency (the Canadian Security Intelligence Service), criminalize the promotion of terrorism, and provide the RCMP with new powers of preventative arrest. It includes the following key features:

- Easing the transfer of information between federal agencies, including confidential data in the hands of Passport Canada and the Canada Revenue Agency, to "better detect and act upon threats." The measure applies to activities that "undermine the security of Canada," while granting a specific exemption for lawful advocacy, protest, and dissent;
- Amending the Secure Air Travel Act to make it easier for authorities to deny boarding on a plane to a would-be traveler heading to Syria to join Islamic State militants;
- Amending the Criminal Code, by making it easier for police to make preventative arrests and criminalizing the promotion of terrorism;
- Giving new powers to the Canadian Security Intelligence Service to disrupt threats, such as providing more information to a would-be terrorist's family and friends, interfering with a would-be terrorist's travel plans, or intercepting weapons intended for terrorist use. The new powers would be subject to a judicial warrant, along with ministerial approval, in the more extreme cases; and

- Amending the Immigration and Refugee Protection Act by making it easier to protect classified information in immigration proceedings, including attempts to remove noncitizens on security grounds.[5]

The C-51 bill has been approved by Canada's House of Commons and Senate and is now law, however, with the recent election of Justin Trudeau as Prime Minister of Canada, amendments are expected.

Legal Framework and Counterterrorism Finance Regime

Canada has strong legal measures to pursue financial crimes and a rigorous detection and monitoring process to identify money laundering and terrorist financing activities. The Proceeds of Crime (Money Laundering) and Terrorist Financing Act of 2001 includes specific measures to detect and deter money laundering and the financing of terrorism to facilitate the investigation or prosecution of these offenses, including:

- Establishing record keeping and client identification requirements for financial services providers and other persons that engage in businesses, professions, or activities that are susceptible to being used for money laundering, and the financing of terrorist activities;
- Requiring the reporting of suspicious financial transactions and of cross-border movements of currency and monetary instruments; and
- Establishing an agency that is responsible for dealing with reported and other information, Financial Transaction Reports Analysis Centre of Canada (FINTRAC).[6]

The following government agencies in Canada are responsible for addressing the threat of money laundering and terrorist financing:

- FINTRAC
- Canadian Security Intelligence Service
- RCMP
- Canadian Revenue Agency

FINTRAC, created in 2000, is Canada's financial intelligence unit, responsible for detecting, preventing, and deterring money laundering and financing of terrorist activities. From 1 April 2012 to 31 March 2013, FINTRAC made 157 terrorist finance and security threat-related reports to law enforcement and national security partners, up from 116 the prior year. FINTRAC made 1,143 disclosures to law enforcement and other government agencies from 1 April 2013 to 31 March 2014. Of these, 845 disclosures were money laundering related, 234 were terrorism financing or security threat related, and 64 were both money laundering and terrorism financing or security related. Though the legislative framework does not allow law enforcement agencies direct access to FINTRAC's databases, FINTRAC can share actionable financial intelligence to assist money laundering and terrorist financing investigations. When FINTRAC has determined reasonable grounds exist to suspect information would be relevant to an investigation or prosecution of a money laundering/terrorist financing offense, they are required to disclose financial intelligence to the appropriate authorities.[7]

Canada has criminalized terrorist financing in accordance with international standards; freezes and confiscates terrorist assets without delay; monitors and regulates money/value transfer and other remittance services; requires collection of data for wire transfers; obligates nonprofits to file suspicious transaction reports and monitors them to prevent misuse/terrorist financing; and routinely distributes UN lists of designated terrorists and terrorist organizations to financial institutions.[8]

In June 2014, Canada became one of the first countries to implement comprehensive legislation regulating virtual currencies, such as Bitcoin. The new regulations include provisions that subject digital currencies to the same reporting requirements as money services businesses like Western Union, including the requirement that all digital currency exchanges register with FINTRAC. Digital currency exchanges also will be subject to verification, registration, and record-keeping requirements, including the obligations to report suspicious transactions, implement compliance programs, and determine if any of their customers are politically exposed persons.[9] The new legislation also covers foreign companies that have a place of business in Canada and those directing services at Canadians. Financial institutions will be prohibited from establishing and maintaining bank accounts for customers involved with virtual currency businesses that are not registered under FINTRAC.[10] This measure illustrates Canada's attempt to keep up with financial innovations like virtual currencies that could be used to launder funds or finance terrorism.

International Cooperation

Canada is a trusted partner in international CT efforts. On the CT financing front, it is a member of the Financial Action Task Force (FATF), the Egmont Group, the Asia/Pacific Group on Money Laundering, and is a supporting nation of the Caribbean FATF. Canada is also an observer in the Council of Europe's Select Committee of Experts on the Evaluation of Anti-Money Laundering Measures and the FATF of South America against Money Laundering.[11]

Canada and the United States maintain a close, cooperative CT partnership, working together on key bilateral homeland security programs such as the Beyond the Border initiative and the Cross Border Crime Forum. Canada has supported global efforts to prevent radicalization, counter violent extremism, and promote the rule of law overseas. It has made significant contributions to the Global Coalition to Counter ISIL. In addition to providing military forces to coalition air and ground operations in Iraq, Canadian law enforcement and security services are working to prevent the flow of foreign terrorist fighters to and from Iraq and Syria. Traveling abroad to commit acts of terrorism is a violation of Canadian federal law. Measures include denial of passport applications (or revocation of valid passports) of Canadian citizens suspected of traveling abroad (or aspiring to travel abroad) to commit acts of terrorism, and maintenance of a watch list of individuals (both citizens and noncitizen residents) flagged for potential involvement with violent extremist organizations.[12]

Mexico

As a major drug producing and transit country, Mexico is primarily focused on national security threats posed by TCOs, such as the Sinaloa cartel and other drug trafficking organizations, rather than terrorism. Nevertheless, the Mexican government has remained on alert for possible domestic and international terrorist threats and closely cooperates with the U.S. on bilateral security issues. There has been speculation for years that terrorists or weapons of mass destruction could enter the U.S. by crossing the southern border from Mexico.

In hearings before the Senate and House Armed Services Committees in 2014, General Kelly said that budgets cuts are "severely degrading" the military's ability to defend southern approaches to the U.S. border. He warned that neglect has created vulnerabilities that can be exploited by terrorist groups, describing a "crime-terror convergence" already seen in Lebanese Hezbollah's involvement in the region.[13] In his March 2015 Congressional testimony, General Kelly said, "in addition to thousands of Central Americans fleeing poverty and violence, foreign nationals from countries like Somalia, Bangladesh, Lebanon, and Pakistan are using the region's human smuggling networks to enter the United States. While many are merely seeking economic opportunity or fleeing war, a small subset could potentially be seeking to do us harm."[14] He also expressed the following concern regarding ISIL foreign fighters: "With little ability to track and monitor foreign fighters when they return, it would be relatively easy for those fighters to 'walk' north to the U.S. border along the same networks used to traffic drugs and humans."[15]

According to the State Department, there were no known international terrorist organizations operating in Mexico, despite several erroneous press reports to the contrary during 2014. There was no evidence that any terrorist group has targeted any U.S. citizens in Mexican territory. The Mexican government has strengthened its legal framework to address acts of terrorism, including terrorist financing, and cooperates closely with relevant U.S. Government agencies on third-country nationals who may raise terrorism concerns.[16]

Legal Framework and Counterterrorism Finance Regime

Mexico has undertaken several measures to enhance its CT regime, particularly on the financial front. On 11 February 2014, the Mexican Senate approved amendments to the Federal Penal Code, the Federal Criminal Procedure Code, the Organized Crime Law, the Federal Fiscal Code, the Asset Forfeiture Law, and Constitutional implementing legislation. These amendments strengthened Mexico's legal framework to address acts of terrorism, terrorist financing, and third-party assistance to the financing of terrorism, attacks against internationally protected persons, the conspiracy to commit terrorism, theft of radioactive or nuclear materials, and the sanctioning of the freezing or forfeiture of terrorist assets based on domestic and international intelligence sources. Minimum sentences for acts of terrorism were increased from 6 to 40 years to 15 to 40 years, the penalties for crimes committed using illicit resources were strengthened, and an exception to rules governing the dissemination of third-party fiscal data in order to comply with new terrorist financing laws was created.[17] In October 2012, Mexico's president signed long-awaited anti-money laundering legislation into law. As a result, the Federal Law for the Prevention and Identification of Operations with Illicit Resources, which went into effect on 17 July 2013, targets "vulnerable" transactions or activities that could be exploited for money laundering and terrorist financing.[18]

The following government agencies in Mexico are responsible for addressing the threat of money laundering and terrorist financing:

- Financial Intelligence Unit
- Ministry of Finance and Public Credit
- Attorney General's Office
- Ministry of Foreign Affairs
- National Insurance and Bond Commission
- National Banking and Securities Commission
- Tax Administration Service

Mexico can still expand on its capacity to proactively investigate and detect terrorism-related activities. Specialized units exist within the Mexican Attorney General's Office that focus on organized crime and money laundering, but Mexican authorities could improve cooperation with other government entities, such as the Mexican Finance Secretariat's Financial Intelligence Unit. In January 2014, the head of the Mexican Financial Intelligence Unit publicly disseminated rules outlining its power to order financial institutions to freeze the assets of designated persons and entities, namely those involved in illicit proceeds. In August 2014, rules limiting individual and business deposits in banks were changed. Previously, banks could not accept more than $4,000 per month from an individual account holder, or more than $14,000 from business entities operating in the U.S. border region or defined tourist areas. The changes allow border and tourist area businesses to exceed the $14,000 per month cash deposit limit provided that they: 1) have been operating for at least three years; 2) provide additional information to financial institutions justifying the need to conduct transactions in U.S. dollars cash; and 3) provide two years of financial statements and tax returns.

The private sector has also taken steps to mitigate the risks associated with terrorist financing and money laundering in Mexico. JP Morgan, Bank of America, and Citigroup-owned Banamex USA have shut four branches in the border town Nogales, almost halving the number in that town owned by big U.S. banks, in the past several months. Separately, hundreds of Chase and Wells Fargo customers, some of them second- and third-generation business owners, have had their bank accounts closed. These bank moves come amid a recent industry-wide focus on enhancing anti-money laundering and CT financing. Wall Street firms want to avoid the huge fines that could result if firms are drawn into the flow of dirty money, but these measures are making it more difficult for legitimate cross-border businesses to operate in the international financial system.[19]

Mexico is focused on countering TOC as it is a major drug producing and transit country. Proceeds from the illicit drug trade leaving the United States are the principal source of funds laundered through the Mexican financial system. Other significant sources of laundered funds include corruption, kidnapping, extortion, intellectual property rights violations, human trafficking, and trafficking in firearms. Sophisticated and well-organized drug trafficking organizations based in Mexico take advantage of the extensive U.S.-Mexico border, the large flow of legitimate remittances, Mexico's proximity to Central American countries, and the high volume of legal commerce to conceal illicit transfers to Mexico. The smuggling of bulk shipments of U.S. currency into Mexico and the repatriation of the funds into the United States via couriers or armored vehicles remains a commonly employed money laundering technique.

Additionally, the proceeds of Mexican drug trafficking organizations are laundered using variations on trade-based methods, particularly after Mexico put restrictions on U.S. dollar deposits. For example, checks and wires from so-called 'funnel accounts' are used by Mexico-based money 'brokers' to acquire goods which are exchanged for pesos in Mexico, or to sell dollars to Mexican businesses. Many of these money laundering risks and methods can be used in the financing of terrorism. The combination of a sophisticated financial sector and a large cash-based informal sector complicates money laundering and terrorist financing countermeasures.[20] Over the past few years, Mexico has significantly enhanced its anti-money laundering and counterterrorist financing capabilities to safeguard its financial system, promote transparency, and attract more foreign direct investment.

International Cooperation

Mexico is a member of the FATF, an observer of the Council of Europe Committee of Experts on the Evaluation of Anti-Money Laundering Measures and the Financing of Terrorism, and a non-observer special status member of the Caribbean FATF. On broader CT issues, Mexico continues to work with the Organization of American States (OAS)/Inter-American Committee Against Terrorism (CICTE) to implement a joint CT work plan, which includes nonproliferation and weapons of mass destruction interdiction. OAS/CICTE collaborated closely with the Export Control and Related Border Security Program on this initiative, and in 2013, the committee funded multiple CICTE workshops in Mexico City focused on building awareness and best practices. In May 2014, Mexico hosted the Global Initiative to Combat Nuclear Terrorism plenary in Mexico City.[21]

United States

The tragic terrorist attacks of 11 September 2001 perpetrated by al-Qaeda definitively transformed U.S. national security and established robust CT measures at home and abroad. The historic attack on the homeland led to the establishment of the DHS, U.S. Northern Command, the Office of the Director of National Intelligence, and the military campaign known as Operation Enduring Freedom to pursue al-Qaeda in Afghanistan. It also prompted the passage of the USA PATRIOT Act that expanded U.S. intelligence and law enforcement capabilities to combat terrorism. The landmark legislation included titles that enhanced domestic security against terrorism, intelligence collection, surveillance procedures, anti-money laundering measures to prevent terrorism, border security, and information sharing for critical infrastructure protection. Financial intelligence and the financial front against terrorism became integral components of the U.S. CT campaign.

Legal Framework and Counterterrorism Finance Regime

The U.S. CT finance strategy is based on three pillars: law enforcement and intelligence operations, financial regulatory measures, and international engagement. The U.S. CT finance regime expanded on many of the already existing CT and anti-money laundering measures and investigation capabilities prior to 9/11. The Department of Justice is the principal government entity responsible for the investigation and prosecution of terrorist financing offenses at the federal level. It uses its authorities to investigate and dismantle terrorist financiers and deter future supporters. For this mission and recognizing the importance of tracking the financial support for terrorist activity, the Federal Bureau of Investigation (FBI)-Terrorist Financing Operations Section (TFOS) was established immediately after 9/11 to identify and disrupt all terrorist financing activities. TFOS works closely with FBI Joint Terrorism Task Forces. FBI-TFOS is charged with managing the FBI's investigative efforts into terrorist facilitators and ensuring financial investigative techniques are used, where appropriate, in all FBI CT investigations to enhance the investigations.[22]

The DEA's drug trafficking and money laundering enforcement initiatives seek to deny drug trafficking and money laundering routes to terrorist organizations; the Bureau of Alcohol, Tobacco, Firearms, and Explosives investigates the illegal sale of explosives and tobacco products that may have connections with terrorism.[23] Since 2001, more than 229 cases have either led to convictions or are still pending judgment against individuals who were charged with supporting—or conspiring to support—terrorism or terrorist groups through material support, transmitting money without a license, narco-terrorism, and economic sanctions violations.[24]

Within the DHS, Customs and Border Protection detects the movement of bulk cash across U.S. borders and maintains data about the movement of commodities and persons in and out of the United States, while Immigration and Customs Enforcement-Homeland Security Investigations initiates investigations of terrorist financing involving transnational crimes to include smuggling and trade-based money laundering. The Internal Revenue Service (IRS), a bureau within Treasury which administers and enforces U.S. tax laws, also plays a supporting role in the U.S. Government's counterterrorist financing efforts, in particular through the work of IRS-Criminal Investigations, which investigates criminal violations of U.S. tax law, as well as money laundering and other financial crimes, and IRS Tax Exempt and Government Entities Division, which administers IRS regulations related to tax-exempt charitable organizations.[25]

Since 9/11, the U.S. Government focused increasingly on the importance of disrupting the finances and funding networks that support terrorist organizations and on the importance of financial intelligence collected by domestic financial institutions. The Department of Treasury's Office of Terrorism and Financial Intelligence (TFI) was established in 2004 to lead the U.S. Government's CT finance efforts. TFI seeks to mitigate the risk of terrorist financing through both systemic and targeted actions. Targeted actions, usually in the form of targeted financial sanctions administered and enforced by Office of Foreign Assets Control, are used to identify, disrupt, and prevent terrorists from accessing the U.S. financial system.[26]

These actions are complemented by the efforts of Financial Crimes Enforcement Network and the federal bank regulators, like the Federal Reserve or Office of the Comptroller of the Currency, that evaluate and enforce a financial institution's compliance with the appropriate regulatory requirements. For example, as administrator of the Bank Secrecy Act, Financial Crimes Enforcement Network, a component of TFI, issues implementing regulations for the Bank Secrecy Act to reduce the potential for abuse by various illicit finance threats including terrorist financing.

The following government agencies are responsible for addressing the threat of money laundering and terrorist financing in the U.S.:

- Department of the Treasury
    - Terrorism and Financing Intelligence
    - Financial Crimes Enforcement Network
    - Office of Foreign Assets Control
    - Office of Intelligence and Analysis
    - Treasury Executive Office of Asset Forfeiture
    - Internal Revenue Service
    - Criminal Investigation
    - Tax Exempt and Government Entities Division
    - Small Business/Self Employed Division
- Department of Justice
    - Federal Bureau of Investigation-Terrorist Financing Operations Section
    - National Security Division
    - Tax Division
    - Drug Enforcement Administration
- Department of Homeland Security
    - Immigration and Customs Enforcement, Homeland Security Investigations
    - Customs and Border Protection

    - Office of Intelligence and Analysis
- Department of State
    - Bureau of Counterterrorism
    - Bureau of Economic and Business Affairs
    - Bureau of International Narcotics and Law Enforcement
- National Counterterrorism Center
- Staff of the federal bank regulators
- Department of Defense
- Combatant Commands
- Defense Intelligence Agency

The DOD has multiple supporting roles in the CTF arena. CTF refers to the activities and actions taken by U.S. Government agencies to deny, disrupt, destroy, or defeat threat finance systems and networks that provide financial and material support to terrorists, insurgents, drug traffickers, weapon traffickers, human traffickers, or corrupt government officials. The use of financial intelligence as a methodology to identify and disrupt terrorist organizations has significantly increased across the intelligence community. The senior DOD leadership has recognized the significance, both strategically and tactically, of a capable and robust CTF posture with the following responsibilities:

- The Under Secretary of Defense, Intelligence is responsible for providing the DOD contribution to foreign intelligence and counterintelligence as part of the Defense Intelligence Agency mission;
- United States Special Operations Command (USSOCOM) has been charged with synchronizing the CT plans of the six geographic combatant commands; and
- Combatant commands have been charged with planning, executing, and synchronizing day-to-day CTF activities within their respective areas of responsibility or functional areas.[27]

International Cooperation

The U.S. is a leader in global CT efforts at the UN and other multilateral venues, including on the financial front. Recognizing the interconnectedness of the global financial markets and their vulnerability to financial crimes like terrorist financing, the U.S. stresses the importance of safeguarding the U.S. and international financial system. The U.S. Government engages bilaterally and multilaterally to globalize its CT finance efforts through the following initiatives:

- Support the development of strong international anti-money laundering/countering the financing of terrorism standards and work toward robust implementation of them through the FATF and the UN as well as other bodies;
- Raise international awareness of the nature and characteristics of terrorist financing as well as calling attention to specific threats; and
- Providing training and technical assistance to bolster national countering the financing of terrorism regimes and enforcement mechanisms. Help to strengthen global anti-money laundering/counter the financing of terrorism regimes directly benefits the safety and integrity of the U.S. financial system, given the global nature of money laundering and threat finance, and the relationships between banks abroad.[28]

Recent Efforts to Combat Terrorism and its Financing in North America

The significant military advances in Iraq and Syria by ISIL, its brutal massacres, and its foreign fighter recruitment have revitalized CT efforts around the globe, including in the U.S., Canada, and Mexico. One year ago, ISIL proclaimed its Islamic Caliphate and is considered the richest terrorist group in the world. To complement its regional aspirations in the Mideast, ISIL has conducted an aggressive social media campaign to instill terror by broadcasting its atrocities and to inspire and recruit sympathizers in the West. According to U.S. Director of National Intelligence James Clapper's testimony before Congress in February 2015, ISIL has recruited over 20,000 foreign fighters, including 3,400 from the West, with about 600 from the United Kingdom and 150 from the United States.[29]

In response to the rise of ISIL, the U.S. has built a global coalition of willing partners (including Canada and Mexico) with the goal of degrading and ultimately defeating ISIL. President Obama set forward a comprehensive strategy featuring nine lines of effort to counter ISIL:

1. **Supporting Effective Governance in Iraq.** We are supporting the new Iraqi government on efforts to govern inclusively and effectively as well as to strengthen its cooperation with regional partners.
2. **Denying ISIL Safe-Haven.** We are conducting a systematic campaign of airstrikes against ISIL in Iraq and Syria. Working with the Iraqi government, we are striking ISIL targets and supporting Iraqi forces on the ground. We will degrade ISIL's leadership, logistical and operational capability, and deny it sanctuary and resources to plan, prepare, and execute attacks.
3. **Building Partner Capacity.** We will build the capability and capacity of our partners in the region to sustain an effective long-term campaign against ISIL. Our advisors are working to advise Iraqi forces, including Kurdish forces, to improve their ability to plan, lead, and conduct operations against ISIL, and we will provide training to help the Iraqis reconstitute their security forces and establish a National Guard. Our train and equip program will strengthen the Syrian moderate opposition and help defend territory from ISIL.
4. **Enhancing Intelligence Collection on ISIL.** Continuing to gain more fidelity on ISIL's capabilities, plans, and intentions is central to our strategy to degrade and ultimately destroy the group, and we will continue to strengthen our ability to understand this threat, as well as to share vital information with our Iraqi and Coalition partners to enable them to effectively counter ISIL.
5. **Disrupting ISIL's Finances.** ISIL's expansion over the past year has given it access to significant and diverse sources of funding. So, we are working aggressively with our partners on a coordinated approach to reduce ISIL's revenue from oil and assets it has plundered, limit ISIL's ability to extort local populations, stem ISIL's gains from kidnapping for ransom, and disrupt the flow of external donations to the group.
6. **Exposing ISIL's True Nature.** Clerics around the world have spoken up to highlight ISIL's hypocrisy, condemning the group's savagery and criticizing its self-proclaimed "caliphate." We are working with our partners throughout the Muslim world to highlight ISIL's hypocrisy and counter its false claims of acting in the name of religion.

7. **Disrupting the Flow of Foreign Fighters.** Foreign terrorist fighters are ISIL's lifeblood, and a global security threat—with citizens of nearly 80 countries filling its ranks. On 24 September, the president convened an historic Summit-level meeting of the UN Security Council, focused on this issue and we will continue to lead an international effort to stem the flow of fighters into Syria and Iraq.

8. **Protecting the Homeland.** We will continue to use the criminal justice system as a critical counterterrorism tool, work with air carriers to implement responsible threat-based security and screening requirements, and counter violent extremism here at home.

9. **Humanitarian Support.** We and our partners will continue to provide humanitarian assistance to the displaced and vulnerable in Iraq and Syria.

Pursuing these lines of effort to advance the comprehensive strategy is a whole-of-government effort.[30]

While much of the fight against ISIL has focused on the military air campaign Operation Inherent Resolve over Iraq and Syria, North American law enforcement and intelligence agencies have stepped up efforts to protect the homeland from homegrown terrorists inspired by ISIL and prevent sympathizers from joining and supporting ISIL. In February 2015, FBI Director James Comey revealed that his agency was investigating suspected supporters of ISIL in various stages of radicalizing in all 50 states.[31] In 2015, U.S. authorities have arrested scores of suspected ISIL sympathizers accused of providing material support of terrorism, joining ISIL, or recruiting for the group. According to a study from Fordham University Law School, federal prosecutors have charged 56 people for supporting ISIS since March 2014; law enforcement killed three other suspects. Fordham researchers say most of the accused are U.S. citizens with more than 60 percent of those charged 21 years old or younger, and more than 80 percent of the cases involved recruitment via social media.[32]

Fears of a 4 July 2015 Independence Day terrorist attack on the homeland resulted in the most dramatic deployment of security forces at the federal, state, and local levels across the United States since 9/11. The DHS publicly warned of an increased threat of ISIL-inspired violence over the holiday weekend. But, there was no "specific, credible intelligence" about an attack, only that ISIL had called for attacks "against members of the military, law enforcement, the U.S. Government and the American public" during the holy month of Ramadan that ended in mid-July. New York City deployed some 7,000 additional officers on the streets for the Independence Day festivities, and the FBI set up coordination centers around the country. Law enforcement officials urged all citizens out and about: "If you see something, say something."[33] In a 9 July briefing with reporters, Comey said that more than 10 people inspired by Islamic State militants were arrested in suspected terrorism-related plots during the four to six weeks leading up to Independence Day. He also believed their work disrupted efforts to kill people, likely in connection with the holiday. Comey said the arrests underscored ISIL's ability to use the Internet in "recruiting, directing and motivating" people to carry out terrorist attacks in the United States.[34]

To counter ISIL, one of the nine lines of effort of the U.S. strategy is disrupting its finances. It is focused on disrupting its revenue streams, restricting its access to the international financial system, and targeting ISIL leaders and facilitators with sanctions.[35] The Canadian, Mexican, and U.S. CT finance regimes are responsible for this mission. On the financing front, U.S. law enforcement agencies have identified isolated cases of U.S. persons who have provided or attempted to provide funds to ISIL, as well as U.S. persons who have traveled or attempted to travel overseas to serve as foreign terrorist fighters with or in support of ISIL.[36] On 11 June 2015, a 17-year-old from Virginia pleaded guilty in court to charges of conspiring to help ISIL militants; this was the first time the U.S. has prosecuted a minor as an adult in such a case. Ali Amin, of Manassas, Virginia, used Twitter and his blog to provide instructions on how to use the virtual currency Bitcoin to send funds to the militants, according to court documents. Prosecutors said Amin also helped another Virginia resident, Reza Niknejad, to travel to Syria to join the group.[37] Although this is not an actual case of fundraising via Bitcoin, it does illustrate how ISIL militants and their supporters are active in North America and are looking at new financial innovations as possible vehicles for raising and moving money to support their terrorist activities.

Given heightened concerns over the threat of terrorism, the Canadian legislature has conducted a series of hearings to analyze and recommend measures to strengthen CT measures. On 24 February 2015, the House of Commons Standing Committee on Finance adopted the following motion:

> That the Committee, at the request of the Minister of Finance, undertake a study of the costs, economic impact, frequency and best practices to address the issue of terrorist financing both here in Canada and abroad.[38]

The June 2015 Report of the Standing Senate Committee on Banking, Trade, and Commerce recognized the pervasive threat of terrorism to Canada and recommended:

1. The federal government continue to educate and train legislators, law enforcement agencies and the public about the connection between terrorist financing and terrorist activity. Moreover, any federal actions to fight terrorism should consider the financing angle.
2. The federal government, in its fight against the Islamic State of Iraq and the Levant (ISIL), explore new ways to disrupt ISIL's financing sources. In particular, in addition to building and training local and regional security forces, the government should target ISIL's administrators, financial collectors and distributors.
3. The federal government, in light of the numerous global cases of charities being used to raise and transfer funds for terrorist financing purposes, continue its efforts to bring increased transparency to the charitable sector in Canada.
4. The federal government, with its international allies, track key facilitators of terrorist financing and work with social media networks to recognize, and take action, when their platforms are being used for illegal activities.
5. The federal government work with all relevant stakeholders to create the expertise and operational capabilities that would enable Canada to take a leadership role in counterterrorist financing. Actions in this regard could include supporting a private sector-led financial crime center to act as a center of excellence and to facilitate dialogue between the public and private sectors.
6. The Financial Transactions and Reports Analysis Centre of Canada (FINTRAC) and entities required to report under the Proceeds of

Crime (Money Laundering) and Terrorist Financing Act work to improve communication between them.

7. The federal government, in recognizing that many transfers occur below the current $10,000 threshold, consider lowering the threshold for reporting international electronic funds transfers.
8. The federal government work with the appropriate stakeholders to develop a digital counterterrorism strategy with a view to keeping pace with illicit fundraising using digital technologies.[39]

Conclusion

Money serves as the oxygen for any activity, licit or illicit; it is the lifeblood for any organization including terrorist groups like ISIL. Financial intelligence and investigative tools such as 'following the money trail' are instrumental to better understand, detect, disrupt, and dismantle terror networks. Tracking how terrorists raise, move, store, and use money has been instrumental in degrading and defeating these groups. Since the tragic attacks of 9/11, Canada, Mexico, and the U.S. have incorporated the financial instrument of national power in efforts to combat terrorism and crime. The financial instrument of national power in the CTF arena has been manifested threefold through:

1. Intelligence and law enforcement operations to pursue terrorist financiers and money launderers;
2. Public designations, sanctions, and asset freezes and seizures; and
3. Domestic and international capacity building in the CTF discipline and international cooperation.

Enhanced anti-money laundering and CTF measures have significantly damaged the illicit networks. Over the past decade, al-Qaeda operatives and affiliates from Iraq to Afghanistan complained about increased difficulty in funding terrorist operations, recruiting foreign fighters, and supporting their networks. Similarly, TCOs in the Western Hemisphere, like the Mexican cartels, realized that greater oversight of international bank transactions and offshore accounts post-9/11 undermined their ability to launder profits through the formal banking sector. Following the money trail and the surveillance of facilitators, like the bankers and lawyers moving and sheltering money for terrorist and criminal groups, produced critical financial intelligence that has led to the destabilization of illicit actors such as al-Qaeda and the drug cartels.[40]

Once the tighter measures to fight money laundering and terrorist financing were put into practice after 9/11, they had an unexpected but constructive side effect—rooting out corruption. Mexico strengthened its anti-money laundering regime to track and combat the Mexican drug cartels. Some of these 'follow the money' measures, including tracking suspicious bank transactions, resulted in the 26 February 2013 arrest of the most prominent teacher union leader in Mexico, Edna Esther Gordillo, on corruption and embezzlement charges. Investigators from Mexico's treasury found that more than $200 million had been diverted from union funds into private bank accounts abroad (including Gordillo's) between 2008 and 2012.[41] Gordillo was living large with significant real estate holdings in Mexico City as well as two luxury properties in Coronado, California.[42] The Gordillo case illustrates how financial forensics, intended to pursue terrorists and drug traffickers, are yielding promising corollary results in the fight against corruption.

As described above, Canada, Mexico, and the U.S. have developed robust legal authorities and investigative mechanisms to counter the financing of terrorism. North Americans countering the financing of terrorism regimes are among the most advanced in the world that other countries seek to emulate. In the face of emerging threats like ISIL-inspired homegrown terrorism and foreign fighter recruits in 2014-2015, we have witnessed rapid responses by the executive, legislative, and judicial branches in North America to provide law enforcement and intelligence agencies the authorities and resources necessary to protect against terrorism; but this campaign against terrorism will require sustained interagency and international efforts for years. In combating terrorist financing, the international community must collaborate and leverage all the instruments of national power to dismantle, degrade, disrupt, and deter illicit networks, as money knows no borders. These instruments include diplomatic, military, intelligence, information, law enforcement, economic, and financial tools that can be applied alone or in combination to counter terrorism and other national security threats.

The June 2015 U.S. National Military Strategy states that international efforts to counter violent extremist organizations must disrupt their planning and operations, degrade support structures, remove leadership, interdict finances, impede the flow of foreign fighters, counter malign influences, liberate captured territory, and ultimately defeat them.[43] Such countermeasures have been successfully leveraged at the local, national, and international levels to combat and degrade terror networks around the globe like al-Qaeda, the Tamil Tigers, and the FARC in Colombia; however, to confront new threats like ISIL, these CT strategies and policies must be continually assessed and updated to keep up with the resourcefulness of terrorist groups that adapt to and circumvent our countermeasures, particularly on the financial front.

Endnotes

1. The views expressed in this chapter as those of the author and do not necessarily reflect the views of the William J. Perry Center for Hemispheric Defense Studies, National Defense University, or the Department of Defense. Many thanks to Rebecca Bock and Farouq Ghazzawi who provided research support for this chapter.
2. National Consortium for the Study of Terrorism and Responses to Terrorism (START), accessed at: http://www.start.umd.edu/.
3. U.S. Department of Defense Operation Inherent Resolve website, accessed at: http://www.defense.gov/home/features/2014/0814_iraq/.
4. Steven Chase, "Prime Minister labels shootings as 'terrorist' acts," *The Globe and Mail*, 22 October 2014, accessed at: http://www.theglobeandmail.com/news/national/in-national-address-prime-minister-labels-shootings-as-terrorist-acts/article21252133/#dashboard/follows/.
5. Daniel Leblanc and Chris Hannay, "Privacy, security and terrorism: Everything you need to know about Bill C-51," *The Globe and Mail*, 10 March 2015, accessed at: http://www.theglobeandmail.com/news/politics/privacy-security-and-terrorism-everything-you-need-to-know-about-bill-c-51/article23383976/.
6. Proceeds of Crime (Money Laundering) and Terrorist Financing Act (PCMLTFA) Summary, accessed at: http://www.fintrac-canafe.gc.ca/act-loi/1-eng.asp.
7. State Department International Narcotics Control Strategy Report 2015, accessed at: http://www.state.gov/j/inl/rls/nrcrpt/2015/vol2/239063.htm.
8. Ibid.
9. According to the Financial Action Task Force, a politically exposed person is an individual who is or has been entrusted with a prominent function. Many politically exposed persons hold positions that can be abused for the purpose of laundering illicit funds or other predicate offenses such as corruption or bribery.
10. State Department International Narcotics Control Strategy Report 2015, accessed at: http://www.state.gov/j/inl/rls/nrcrpt/2015/vol2/239063.htm.

11. State Department 2013 Country Reports on Terrorism, Chapter 2 Western Hemisphere Overview, accessed at: http://www.state.gov/j/ct/rls/crt/2013/224825.htm.
12. State Department 2014 Country Reports on Terrorism, Chapter 2 Western Hemisphere Overview, accessed at: http://www.state.gov/j/ct/rls/crt/2014/239409.htm.
13. Molly O'Toole, Top General Says Mexico Border Security Now 'Existential' Threat to U.S." *Defense One*, 5 July 2014, accessed at: http://www.defenseone.com/threats/2014/07/top-general-says-mexico-border-security-now-existential-threat-us/87958/.
14. U.S. Southern Command 2015 Posture Statement to Congress, 12 March 2015, accessed at: http://www.southcom.mil/newsroom/Pages/2015-Posture-Statement-to-Congress-.aspx.
15. Kevin Baron, "US Military Concerned ISIS Fighters Returning to Caribbean Could Reach Border," *Defense One*, 12 March 2015, accessed at: http://www.defenseone.com/threats/2015/03/us-military-concerned-isis-fighters-returning-caribbean-could-reach-Border/107421/.
16. Ibid.
17. Ibid.
18. State Department 2013 Country Reports on Terrorism, Chapter 2 Western Hemisphere Overview.
19. Emily Glazer, "Big Banks Shut Border Branches in Effort to Avoid Dirty Money Customers in Nogales, Ariz., Seek New Lenders After Banks Retreat," *The Wall Street Journal*, 25 May 2015, accessed at: http://www.wsj.com/articles/big-banks-shut-border-branches-in-effort-to-avoid-dirty-money-1432598865.
20. State Department International Narcotics Control Strategy Report 2015, accessed at: http://www.state.gov/j/inl/rls/nrcrpt/2015/vol2/239063.htm.
21. Ibid.
22. Office of the Inspector General, Department of Justice, "Audit of the Federal Bureau of Investigation's and the National Security Division's Efforts to Coordinate and Address Terrorist Financing," Audit Report 13-17, March 2013, accessed at: http://www.justice.gov/oig/reports/2013/a1317.pdf.
23. DEA, Department of Justice, "FY 2014 Performance Budget Congressional Submission," accessed at: http://www.justice.gov/jmd/2014justification/pdf/dea-justification.pdf.
24. U.S. Department of Treasury National Terrorist Financing Risk Assessment 2015, accessed at: http://www.treasury.gov/resource-center/terrorist-illicit-finance/Documents/National%20Terrorist%20Financing%20Risk%20Assessment%20%E2%80%93%2006-12-2015.pdf.
25. Ibid.
26. David Cohen, Under Secretary for Terrorism and Financial Intelligence, Department of the Treasury, Testimony before the Senate Committee on Banking, Housing and Urban Affairs "Patterns of Abuse: Assessing Bank Secrecy Act

Compliance and Enforcement," 7 March 2013, accessed at: http://www.treasury.gov/press-center/press-releases/Pages/jl1871.aspx.

27. Joint Chiefs of Staff Commander's Handbook for Counter Threat Finance, September 2011, accessed at: https://info.publicintelligence.net/DoD-Counter-ThreatFinance.pdf.
28. U.S. Department of Treasury National Terrorist Financing Risk Assessment 2015.
29. Robert Windrem, "ISIS By the Numbers: Foreign Fighter Total Keeps Growing," NBC News, 28 February 2015, accessed at: http://www.nbcnews.com/storyline/isis-terror/isis-numbers-foreign-fighter-total-keeps-growing-n314731.
30. The White House, FACT SHEET: The Administration's Strategy to Counter the Islamic State of Iraq and the Levant (ISIL) and the Updated FY 2015 Overseas Contingency Operations Request, 7 November 2014, accessed at: https://www.whitehouse.gov/the-press-office/2014/11/07/fact-sheet-administration-s-strategy-counter-islamic-state-iraq-and-leva.
31. Jesse Byrnes, FBI investigating ISIS suspects in all 50 states," *The Hill*, 25 February 2015, accessed at: http://thehill.com/blogs/blog-briefing-room/233832-fbi-investigating-isis-suspects-in-all-50-states.
32. PBS News Hour, "Study: Arrests for alleged ISIS supporters on the rise in the U.S.," accessed at: http://www.pbs.org/newshour/bb/arrests-alleged-isis-supporters-grow-accused-common/.
33. "U.S. on high alert heading into July 4 weekend," CBS News, 3 July 2015, accessed at: www.cbsnews.com/news/fourth-july-terror-threat-isis-inspired-attack-keeps-officers-high-alert/.
34. Richard Serrano, "FBI chief says agents thwarted possible July 4 terror attacks," *Los Angeles Times*, 9 July 2015, accessed at: http://www.latimes.com/nation/la-na-fbi-terror-arrests-20150709-story.html.
35. Charles Lister, "Cutting off ISIS' Cash Flow," The Brookings Institution, 24 October 2014, accessed at: http://www.brookings.edu/blogs/markaz/posts/2014/10/24-lister-cutting-off-isis-jabhat-al-nusra-cash-flow.
36. Ibid.
37. Julia Edwards, "American teenager pleads guilty to helping Islamic State," *Reuters*, 11 June 2015, accessed at: http://www.reuters.com/article/2015/06/11/us-usa-security-islamicstate-idUSKBN0OR1V520150611.
38. Report of the Standing Committee on Finance James Rajotte Chair, Canadian House of Commons, "Terrorist Financing In Canada And Abroad: Needed Federal Actions," June 2015, accessed at: http://www.parl.gc.ca/Content/HOC/Committee/412/FINA/Reports/RP8048561/412_FINA_Rpt13_PDF/412_FINA_Rpt13-e.pdf.
39. Ibid.
40. Celina B. Realuyo, "Following the Money Trail to Combat Terrorism, Crime and Corruption," U.S. Institute of Peace, International Network for Economics

and Conflict, 15 April 2013, accessed at: http://inec.usip.org/blog/2013/apr/15/following-money-trail-fight-terrorism-crime-and-corruption.

41. Catherine E. Shoichet, "Mexican union leader accused of embezzling millions," CNN.com, 26 February 2013, accessed at: http://www.cnn.com/2013/02/26/world/americas/mexico-union-leader-arrested/index.html.

42. Sandra Dibble & Lily Leung, "Suspicious Home Purchase Went Unnoticed Suspicious Real Estate Sales Tough to Track," UT San Diego, 24 March 2013, accessed at: http://www.utsandiego.com/news/2013/mar/24/tp-suspicious-home-purchase-went-unnoticed/.

43. The National Military Strategy of the United States of America 2015, accessed at: http://www.jcs.mil/Portals/36/Documents/Publications/2015_National_Military_Strategy.pdf.

Chapter 8. Inside Pandora's Box: Foreign Fighters and the Lone Wolf Terrorism Nexus

Colonel Bernd Horn

The attacks on the offices of Charlie Hebdo in Paris, France, on 8 January 2015 and on a cartoon exhibit in Garland, Texas, on 4 May 2015, underline a pair of threats that are, and will continue to be, a significant challenge for national security organizations. Incredibly, they represent in many ways a hidden cancer that can fester and grow within societies without necessarily manifesting any visible symptoms until too late. These threats are foreign fighters and lone wolf terrorism, which especially when combined, can prove devastating to homeland security. Importantly, they are a menace that can easily transcend borders and, as such, necessitate national, regional, and global responses.

Foreign fighters, while not a new phenomenon, have become a much greater concern to governments worldwide. As a result of globalization, terrorist organizations have been increasingly able to use the Internet and social media to attract, seduce, and subsequently radicalize individuals to join 'the cause' and wage jihad, or support other extremist action. Many groups, such as the Islamic State, have been exceptionally successful at attracting large numbers of foreigners to join and fight for their cause. These volunteers who are recruited to terrorist organizations and then become foreign fighters expand the international reach of transnational insurgencies, as well as religious and ideological conflicts. In fact, research studies have shown

---

Colonel Bernd Horn, OMM, MSM, CD, Ph.D., is a retired Regular Force infantry officer who has held key command and staff appointments in the Canadian Armed Forces. He is currently the Director of the CANSOFCOM Professional Development Centre, an appointment he fills as a reservist. Dr. Horn is also an adjunct professor of the Centre for Military and Strategic Studies, University of Calgary, as well as an adjunct professor of history at the Royal Military College of Canada. He is also a Fellow at the Canadian Defence & Foreign Affairs Institute.

that foreign fighters tend to perpetuate the conflict they have joined, are responsible for higher levels of violence, and act as a recruiting vehicle for other nationals.[1]

History has already shown this threat to be real. Terrorist organizations have trained and sent foreign fighters back to their home or other Western countries to conduct attacks, recruit, and/or train others. Moreover, through social media, the Internet, and agents throughout the world, these same extremist organizations have also preached action in the name of jihad for those individuals unable to leave their countries due to an extensive worldwide governmental clamp-down on allowing individuals suspected of leaving the country to become foreign fighters or to engage in terrorist activities. Once blocked from leaving the country to engage in jihad overseas, these individuals may pose an added threat of lone wolf terrorism to the homeland.

Together, foreign fighters and lone wolf terrorism represent a growing threat that has implications for domestic security, as well as regional stability. The solutions are far from simple and require a comprehensive domestic and regional (if not global) approach. As the barbarity and savageness of the ISIS terrorist organization has shown, to name but one threat organization, turning a blind eye to the cancer of foreign fighters and the organizations they support is an approach fraught with peril. Undeniably, foreign fighters and lone wolf terrorism represent a national, as well as a regional skulking threat and as such, they cannot be ignored.

Initially, it is important to delineate what is meant by the term foreign fighter. Dr. David Malet, an internationally recognized expert on the subject from the University of Melbourne, Australia, characterizes foreign fighters as simply, "non-citizens of conflict states who join insurgencies during civil conflict."[2] Similarly, Barak Mendelsohn, a senior fellow at the Foreign Policy Research Institute, describes them as, "Volunteers [who] leave their homes and intervene in a clash taking place in a foreign location."[3]

Professor Thomas Hegghammer from the Norwegian Defence Research Establishment takes a more complex approach. He classifies the foreign fighter as, "an agent who (1) has joined, and operates within the confines of, an insurgency, (2) lacks citizenship of the conflict state or kinship links to its warring factions, (3) lacks affiliation to an official military organization, and (4) is unpaid."[4] His more detailed definition is structured to exclude mercenaries, returning diaspora members, or exiled rebels, who, as he describes, "have a pre-existing stake in the conflict." In addition, he distinguishes foreign fighters from international terrorists, who specialize in out-of-area violence against noncombatants.[5] In sum, all share the same basic tenant—foreign fighters are individuals who leave their home country to participate in conflict in another state.

The desire to travel to foreign lands to fight for a 'righteous cause,' or simply adventure and thrills is not a new phenomenon. Almost every major conflict has attracted foreign fighters (e.g., Spanish Civil War, Israeli War of Independence, Vietnam War, Soviet invasion of Afghanistan). Importantly, the current wave of foreign fighters has triggered concern, new legislation, and increased law enforcement initiatives worldwide, because the new wave of extremist, radicalized volunteers pose a significant threat to global security. Specifically:

1. They perpetuate the conflict in which they are participating;
2. They act as inspiration to other vulnerable individuals;
3. They create a 'blowback effect'—i.e., they return home, or travel to a third country and commit violent acts, often in the name of jihad; and
4. They are more lethal, dangerous, and sophisticated than their domestic counterparts.[6]

These issues highlight valid concerns. Conflicts worldwide have shown that foreign fighters are responsible for higher levels of indiscriminate violence. Analysts and scholars assert this rise is due to their belief that they must fight more aggressively because they perceive that they are in a losing struggle for the very survival of their cause. In addition, they can be increasingly brutal, savage, and indiscriminate because they have no equity or families to protect in the same manner that local insurgents do.[7] Furthermore, their 'apparent' religious zealousness often promotes sectarian violence.[8] Indeed, Dr. Malet observed: "Transnational recruits are responsible for higher levels of violence than are local insurgents and insurgencies that manage to recruit foreign fighters are disproportionately successful as compared to other rebel groups."[9]

Recent events in Iraq and Syria underline these points. The series of beheadings conducted by ISIS were executed by a British foreign fighter. Videos depicting the mass murder of Syrian soldiers taken prisoner by ISIS, the reports of ultimatums to convert to Islam or face death by those who fell under ISIS control, as well as the litany of stories of captured Kurdish and Yazidis villages and the subsequent killings, rape, and kidnapping of men, women, and children, as well as the accounts of brutal imprisonment by hostages who were later ransomed, all speak to the savagery of ISIS. Undeniably, ISIS has been profoundly successful in its military campaigns to defeat rival forces and capture territory. It has also proven to be immensely capable of attracting a large proportion of foreign fighter recruits. In fact, 40 percent of those in ISIS are foreign fighters.[10] Equally disturbing, many analysts believe that up to 80 percent of the foreign fighters traveling to fight in Syria and Iraq aspire to join ISIS.[11]

Incredibly, despite global reaction by most states to stop individuals from leaving their respective countries to become foreign fighters and/or take part in terrorist activities, as well as the U.S.-led coalition targeting ISIS in Iraq and Syria, and the estimated 10,000 deaths ISIS has sustained to date, the flow of foreign fighters has not diminished. Jurgen Stock, the head of Interpol, revealed, "In September 2014 less than 900 foreign terrorist fighters had been identified by Interpol." He continued, "Today, in less than a year, more than 4,000 profiles are available in our database."[12]

The concern with foreign fighters also revolves around their experience and ideological commitment. Many volunteer to fight for what they see as a 'righteous cause.' Although radicalized through the Internet, social media, or local mentors, they are often indoctrinated to a deeper jihadist ideology as a result of their training and combat experience. As Professor Hegghammer explained, "more importantly they empower transnational terrorist groups such as al-Qaida, because volunteering for war is the principal stepping-stone for individual involvement in more extreme forms of militancy."[13] For example, the London 7/7 bombers (7 July 2005) originally traveled to Afghanistan to fight. However, after having attended a training camp in the Federally Administered Tribal Areas (FATA) for a week, their al-Qaeda instructors directed them to take the fight to England, which had catastrophic results for British citizens.[14]

The danger posed by these fighters is beyond theoretical. A research report noted that between 1990 and 2010, one-in-nine returned foreign fighters were involved in domestic plots. The study revealed that "these plots tended to be more effective and lethal, thanks both to the skills learned and the indoctrinated zeal provided at radical training camps."[15] Currently, in the United Kingdom, "more than half of MI5's [national Security Service] anti-terror investigations involve Britons who have traveled to Syria."[16] Clearly, foreign fighters represent a clear and present danger.

So, who exactly are those willing recruits to fight and die for the apparent cause(s) of others? A well researched 2007 New York Police Department (NYPD) report on radicalization concluded:

> The transformation of a Western-based individual to a terrorist is not triggered by oppression, suffering, revenge or desperation. Rather, it is a phenomenon that occurs because the individual is looking for an identity and a cause and unfortunately, often finds them in extremist Islam. There is no useful profile to assist law enforcement or intelligence to predict who will follow this trajectory of radicalization. The radicalization process is accelerating in terms of how long it takes and the individuals are continuing to get younger.[17]

Similarly, political scientist Robert Pape conducted a specific study of suicide terrorism and he concluded:

> Few suicide attackers are social misfits, criminally insane, or professional losers. Most fit a nearly opposite profile: typically they are psychologically normal, have better than average economic prospects for their communities, and are deeply integrated into social networks and emotionally attached to their national communities.[18]

Despite these findings, the NYPD, as well as MI5, both observed that most individuals "had some vulnerability in their background and [that] made them receptive to extremist ideology and that it was always influenced by others."[19] Numerous other studies have shown foreign fighters tended to be impressionable young males who were students or unemployed, lacking purpose and looking for an identity.[20] French authorities categorize volunteers from France as disaffected, aimless, and lacking a sense of identity or belonging. Scholars tend to agree that these characteristics appear to be common across most nationalities and fit with the high number of converts, presumably people who are seeking a greater sense of purpose and meaning in their lives. Alienation from mainstream society also played a central role. The typical age for recruits averaged from 18 to 29 years old, with some as young as 15 to 17 years old. As noted by the NYPD research, the trend since the mid-2000s is of recruits to extremism becoming younger.[21]

The Homegrown Lone Wolf Nexus

Predictably, for Western governments, including the American and Canadian governments, foreign fighters represent a hidden threat, due to a concept that has often been labeled the 'blow-back effect.' Simply put, once foreign fighters return home, or are ordered home by their respective organizations to continue the fight, they represent a cohort that is more experienced, more lethal, and more dangerous, as well as more sophisticated than their domestic counterparts. They now represent a substantive menace, either as a group or as individuals acting in a lone wolf capacity. The danger they pose is not merely theoretical. As mentioned earlier, a research study revealed that between 1990 and 2010, one-in-nine returned foreign fighters were involved in domestic terrorist plots.[22]

Adding to this already explosive situation is the trend toward lone wolf terrorism, where individuals, or at most a pair of individuals, conduct attacks on their own. Richard Fadden, a former director of the CSIS, testified to a Senate Committee that the shift to the more difficult to detect "sole-actor" or "lone wolf" style attacks is a pressing problem for Western counterterrorist agencies. He conceded, "this makes things very complicated for us [CSIS]." He elaborated that, "the lone-wolf approach tends to attract individuals driven by ideology as well as serious personal problems, a combination that makes them more unpredictable."[23] He explained that the larger group activities or plots allowed security agencies more margin for success. After all, for the plotters to achieve their objective of launching an attack, they had a requirement to plan and communicate. In the modern age of instant communication, this requirement provided scope and some possibility of intercepting transmissions. The fact that there were more players and more moving parts also meant that there was greater likelihood of someone making an error. However, when it is only a single conspirator, when there is only one person not talking to anyone, then as Fadden conceded, "you have to be really lucky."[24]

United States Attorney General Eric Holder agreed. He acknowledged, "It's something that frankly keeps me up at night, worrying about the lone wolf or a group of people, a very small group of people, who decide to get arms on their own and do what we saw in France."[25] The problem was also succinctly described by Dennis Blair, a former director of national intelligence. He acknowledged:

> We have a good capability to detect and disrupt these sorts of multipurpose [terrorist] teams that take months to plan, rehearse, fund, provide the logistics support for an attack. But we are not as capable as we should be of carrying out the much more difficult task of detecting these self-radicalized citizens of the United States, Europe, other countries like Nigeria, who are given a very simple mission – with an advanced bomb to carry it out – or who plan their own attacks, inspired by Al Qaeda's message but not directed by Al Qaeda.[26]

Similarly, former National Security Agency (NSA) Director Lieutenant General Michael Hayden also underlined the concern about the 'new' recruits extremist organizations were training and unleashing back on the Western countries. He described:

> These Western recruits were reputed to speak multiple languages. They were technologically savvy. They understood Western culture and knew how to blend in. Some of the recruits were of Pakistani descent and were part of the huge diaspora that now lived in Britain. But others were Caucasian. Al Qaeda was bringing more and more people into the tribal region, people who wouldn't draw undue attention if they were next to you at the passport line at Dulles Airport.[27]

In essence, the recruitment of Western foreign fighters, especially due to their mobility, has created a serious threat for homeland security, particularly when combined with 'lone wolf' terrorism, which is the term coined to describe individuals who commit, or are prepared to commit, violent terrorist acts on their own, external to, although perhaps loosely affiliated with a recognized terrorist group, in support of a particular ideology or movement.[28] Not surprisingly, the difficulty of identifying these individuals, as discussed earlier, has made lone wolf terrorism an increasingly evolving trend in terror tactics.

The lethal effects of lone wolf terrorism were clearly shown in 1995, when Timothy McVeigh killed over 150 people and injured in excess of 500 more in his attack on a federal building in Oklahoma City, as well as in August 2011, when Anders Breivik murdered 77 individuals in a bombing and shooting spree in Norway. It is not surprising then that a 2009 United States DHS assessment concluded that lone wolf terrorists "are the most dangerous domestic terrorism threat in the United States ... because of their low profile and autonomy."[29] According to University of California Los Angeles lone wolf terrorism expert Jeffery D. Simon:

> What makes lone wolves so dangerous is their ability to think outside the box. Since they operate by themselves, there is no group pressure or decision-making process that might stifle creativity. Lone wolves are free to act upon any scenario they can dream up. This freedom has resulted in some of the most imaginative terrorist attacks in history. For example, lone wolves were responsible for the first vehicle bombing (1920), major midair plane bombing (1955), hijacking (1961), and product tampering (1982), as well as the anthrax letter attacks in the United States (2001).[30]

The prospect of foreign fighters returning home or to other Western countries undetected using their Western passports and then carrying out lone wolf attacks has panicked many, including senior politicians. "This [foreign fighter/IS threat] is a turning point in the war on terror," South Carolina Senator Lindsay Graham remarked on Fox News. He demanded the U.S. president to deploy thousands of ground troops to Iraq, "before we all get killed back here at home." He was not alone. "They intend to kill us," warned House Speaker John Boehner, "And if we don't destroy them first, we're going to pay the price."[31]

Their concerns, although on the surface may appear a tad overdramatic, are not merely theoretical or philosophical. They are in fact very real as they have been borne out. Some recent examples make the point. For example, the 7 July 2005 (7/7) suicide bombings in London, England were homegrown attacks. Two of the four 7/7 bombers, all of whom were British Muslims, had trained in the FATA region of Pakistan and rather than fight in Afghanistan as they had originally intended, their al-Qaeda handlers sent them back to London to conduct a series of coordinated suicide attacks in central London that targeted the public transportation system during the morning rush hour. They used organic peroxide-based explosive devices that were packed in rucksacks. Three of the bombs were detonated in London Underground subway trains. The fourth bomb was detonated on a double-decker bus in Tavistock Square. The four bombers killed themselves and 52 civilians, and in excess of 700 others were injured.

British authorities also thwarted a plot that was described as a "Mumbai-style" armed assault in late February 2014. In June of the same year, French authorities arrested a French national who returned from Syria and conducted an attack in Belgium in May, which killed three people at a Jewish museum. French police also thwarted a nail bomb attack near Cannes. And in Kosovo, arrests in November 2013 apparently disrupted a terrorist cell planning the purchase of weapons for future operations.[32]

Australia has also felt the specter of foreign fighters and homegrown activities. David Irvine, the director general of Australia's spy agency, revealed that 15 Australians fighting with militant groups were believed to have been killed in Iraq and Syria. He indicated that dozens of Australian foreign fighters have already returned home and he conceded, "a good number of these" remained a concern to the authorities. He also revealed that 100 or more people in Australia were "actively supporting" militant groups by recruiting new fighters, grooming suicide-bombing candidates, and providing funds and equipment.[33]

The United States has also been impacted by the peril of homegrown attacks. Since 9/11 there have been in excess of 40 terrorist plots in the U.S. involving American citizens or permanent residents.[34] Within a 15 month period, there were 53 indictments for individuals planning or attempting to conduct terrorist activities in the United States.[35] Some examples prove the severity of the threat. Najibullah Zazi, a 24-year-old coffee cart vendor in Manhattan, who later became a shuttle bus driver at Denver international airport, flew to Peshawar in 2008 with two high school friends eager to join the fight in Afghanistan. While in Pakistan, three senior al-Qaeda leaders persuaded them that they could optimize their assistance to the jihadist cause by returning to New York and conducting a terrorist attack. As such, Zazi plotted to explode a suicide bomb in the New York subway system in a coordinated "Martyrdom" attack with two others in September 2009.[36]

In addition, on 5 November 2009, Major Nidal Hasan, a serving member in the U.S. Army, went on a shooting spree on Fort Hood, Texas, killing 13 people. Also, Faisal Shahzad, a financial analyst at Elizabeth Arden, traveled to a Taliban training camp in the FATA where he learned to make bombs. As was the case with others, Taliban leaders requested Shahzad to return to the U.S. to conduct a terrorist attack. As a result, he planted a car bomb in Times Square, New York, on 1 May 2010. Fortuitously, the bomb failed to go off and two alert street vendors noticed smoke coming from a car and upon investigation spotted the bomb and alerted the New York Police Department.[37]

Examples of other American homegrown attacks, however, do not stem from foreign fighters but rather are the product of radicalized individuals who lashed out. Such was the case of the 19-year-old Somali-born U.S. citizen who tried to detonate what he thought was a car bomb at a Christmas tree lighting ceremony in Portland, Oregon, on 26 November 2010. Approximately 10,000 people had gathered for the ceremony and had the FBI not thwarted the plot, many would have been killed or injured.

Other examples include: a Baltimore construction worker who plotted to blow up a military recruiting station in Maryland; a 34-year-old naturalized American born in Pakistan who was charged with plotting to bomb the Washington Metro; and finally the case of Dzhokhar and Tamerlan Tsarnayev, two brothers who lived in the Boston area for several years and placed two homemade pressure cooker bombs at the finish line of the Boston marathon on 15 April 2013, killing three people and injuring more than 200. The subsequent manhunt paralyzed Boston and ended in a gunfight that killed Tamerlan and led to the capture of Dzhokhar.[38] Most recently, as mentioned, on 4 May 2015, two gunmen, who were indirectly linked to the Islamic State through their twitter accounts, were killed when they attempted to storm a cartoon exhibit and contest related to depicting the prophet Muhammad.[39]

Significantly, a recent study conducted by the Southern Poverty Law Center revealed that between the period of 1 April 2009 to 1 February 2015, a domestic terrorist attack, emanating from the radical right and homegrown jihadists, occurred every 34 days on average. It further demonstrated that 74 percent of the more than 60 incidents examined were carried out or planned by a lone wolf working entirely alone, and a total of 90 percent of the events were the work of just one or two people.[40]

Canada has not been immune to the homegrown threat. In 2006, Canadian homegrown terrorists, who were labeled the "Toronto 18" by police, wanted to prod the Canadian government into rethinking its involvement in Afghanistan. They plotted to target the Toronto area by destabilizing the economy through attacks on the Toronto Stock Exchange by way of three truck bombs set off over three consecutive days, which they hoped would paralyze Canadians with fear and keep them at home. They believed that the attacks would be bigger than the London 7/7 subway bombings. In addition, they also planned to attack the CSIS headquarters in Toronto and an unspecified military base off of Highway 401 between Ottawa and Toronto. CSIS and law enforcement discovered the plot through an informant who assisted them in making the case. As such, police found plans and materials at the homes of the plotters. A test of the terrorists' plans demonstrated "the blast effect from the bomb was equivalent to 768 kilograms of TNT, and would have caused catastrophic damage to a multi-story glass and steel frame building 35 meters from the bomb site, as well as killing or causing serious injuries to people in the path of the blast waves and force."[41]

All members of the group were arrested prior to enacting the plan; however, only four were actually charged with the bomb plot itself as the others were not fully implicated in what the four were planning. Notably, the other 14 were charged with a variety of terrorist-related charges. Although the plot was not activated, all was in place with the exception of the last component, the delivery of three tons of highly combustible ammonium nitrate fertilizer, which, unknowingly to the conspirators, was being handled by a Muslim businessman turned informant who was working with CSIS and the RCMP.[42] Upon delivery of the explosive fertilizer the police swooped in and arrested the extremists. Fortunately the plan never came to fruition. The judge hearing the case concluded the plot, "would have resulted in the most horrific crime Canada has ever seen."[43]

More recently, in April 2013, 'Project Smooth Arrests,' resulted in the apprehension of two Canadians who conspired to attack a VIA Rail passenger train traveling between New York and Toronto. Chiheb Esseghaier and Raed Jaser were arrested and charged. Months later, in July 2013, John Nuttall and Amanda Korody were charged with conspiring to use improvised explosives built using pressure cookers to bomb the British Columbia legislature building during Canada Day festivities.[44]

The homegrown threat is exacerbated by the insidious reach of social media and the Internet. Organizations such as ISIS, which employ thousands of foreign fighters that can be unleashed on the West, also employ savvy cyber skills that allow them to penetrate countries from afar. Michel Juneau-Katsuya, a former Canadian intelligence officer who now heads up an Ottawa cybersecurity company observed ISIS, "have been capable (of recruiting) young people without speaking directly to them face to face." He noted, "If they've been capable through the media—through the Internet—to reach them, I wouldn't be surprised that they convince some of these people that instead of traveling to the Middle East, to stay here and simply attack us here."[45]

In fact, the prime minister's warning was only too real. In July 2014, Martin Rouleau, a 25 year-old radicalized Muslim convert, was identified by the RCMP as a "high-risk" traveler. He was arrested at the airport as he was leaving for Turkey and his passport was seized. Later, on 20 October, Rouleau ran over two military personnel with his car as they walked across a parking lot, killing Warrant Officer Patrice Vincent and seriously injuring another service member.

Then, on 22 October 2014, another radicalized Muslim convert, Michael Zehaf-Bibeau, struck in the nation's capital. He first shot to death Corporal Nathan Cirillo, who was standing guard at the National War Memorial. Subsequently, Zehaf-Bibeau left the scene and quickly moved to Parliament Hill where he entered the House of Commons, penetrating deep inside until he reached the party Caucus rooms, which were in session. However, before he could do anymore damage, he was gunned down by Kevin Vickers, the sergeant-at-arms, and unidentified RCMP personnel. The attack put Ottawa on lockdown and prompted the CAF to institute higher security measures since its members had become the target of radicalized individuals.

This threat prompted Prime Minister Stephen Harper to assert with regard to "knife and gun" attacks in the U.S. and Canada, "These threats are very, very real."[46] In fact, RCMP Commissioner Bob Paulson affirmed, "We have 63 active national security investigations on 90 individuals who are related to the travelling group, people who intend to go [to countries such as Iraq and Syria] or people who have returned."[47] Alarmingly, CSIS Director Michael Coloumbe affirmed, "in the last three to four months numbers of Canadians travelling for extremism in Iraq and Syria has increased by 50 percent."[48]

Solving the Foreign Fighter and Lone Wolf Issues

With the clear and present danger that foreign fighters, the organizations they support, and the homegrown plots that they may inspire, support, initiate, or represent, the overriding question then becomes, how does one stop them? Predictably, many Western governments have taken action. For example, the Canadian Government has taken numerous steps toward tackling the problem and its associated issues. The RCMP leads a High Risk Travel

Case Management Group, which involves a number of other government departments and law enforcement agencies. The group examines cases of extremist travelers and works to find the best tailored response to the most pressing cases through a continuum of actions.

The government also works with international organizations, such as the UN, NATO, G-7, the Global Counter-Terrorism Forum, and Interpol, as well as the FATF to counter terrorism. The government has also taken action through the FINTRAC, releasing more than 200 financial intelligence disclosures to authorities relating to terrorist financing.

Moreover, in 2012, then-Public Safety Minister Vic Towes unveiled Canada's first comprehensive CT strategy, which includes response plans in the event of a major attack and strategies for de-radicalizing homegrown terrorists. Known as "Building Resistance Against Terrorism," the strategy encompasses a four-step methodology that includes preventing, detecting, denying, and responding to possible threats.[49] As part of the government's strategy they also implemented Bill S-7, the Combating Terrorism Act, which came into force in July 2013. This legislation created four new offenses intended to prevent and deter persons from leaving Canada for certain terrorism-related purposes. Specifically, an individual commits an offense by leaving or attempting to leave Canada for the purpose of:

1. Knowingly participating in or contributing to any activity of a terrorist group for the purpose of enhancing the ability of any terrorist group to commit a terrorist activity, including providing training, receiving training, or recruiting a person to receive training;
2. Knowingly facilitating a terrorist activity;
3. Committing an indictable offense on behalf of, at the direction of, or in association with a terrorist group; and
4. Committing an indictable offense that constitutes a terrorist activity.

The offenses described in the first bullet carry a maximum penalty of 10 years imprisonment. The remaining offenses carry a maximum penalty of 14 years.[50]

In addition, the Canadian Government tabled Bill C-51, Anti-Terrorism Act, 2015, in January 2015. Its main provisions facilitate information sharing among a large number of federal departments and institutions, provide enhanced police powers that would allow them to detain or restrict terror suspects in a preventative manner, ban the 'promotion of terrorism,' allow the Minister of Public Safety to add individuals to Canada's 'no-fly list,' and enhance the powers of CSIS.

Other governments (e.g., Australia, the United Kingdom, U.S.) have similarly made it a criminal offense to leave the country to engage in terrorist activities or fight in foreign conflicts. In addition, the UN Security Council, at a meeting chaired by President Obama, unanimously passed a resolution on 23 September 2014, specifically designed to plug the flow of foreign fighters to militant organizations. In fact, the Security Council voted 15 to 0 to compel countries to make it a crime for their citizens to travel abroad to fight with militants or recruit other people to do it.[51]

Although the efforts are laudable, there is one major problem; experts point out that repression, criminalization, and prosecution can be counterproductive as these measures increase the "victimization" narrative. The European Union commission established the Radicalisation Awareness Network, which clearly cautioned:

> Only repression ... will not solve the problem. Prevention, signaling and providing programs to help (potential) foreign fighters to leave the path of violent extremism are necessary as well. These actions are often organized on a local level. For instance, first line practitioners such as teachers and youth workers, can be trained to recognize and refer those who are being influenced to go on jihad. Also, families can be partners in both detecting potential fighters and convincing them to deploy their engagement in a non-violent way. Finally, exit-programs that have proven to be effective, can be tailored to the target group, for instance by employing formers or practitioners as acceptable intermediaries or coaches.[52]

The reality is that many of those who become foreign fighters and join extremist groups do so because they feel alienated or disassociated with their society and/or are looking for a meaning/cause to guide their existence.[53] For many, Islam and the call to defend the *umma* (Muslim community) filled that void. As the Radicalisation Awareness Network notes, further sanctions and threats simply reinforce the narrative of Western victimization of Muslims.

Unfortunately, to date the West offers no clear anti-jihadist message to the many who are drawn to the call to defend Islam. One study identified a wide range of motivations, including, "the horrific images of the conflict, stories about atrocities committed by Governmental forces, and the perceived lack of support from Western and Arab countries." As one analyst noted, "Such motivations not only speak to fellow Muslims, but also to secular-minded people who wish to defend the widow and the orphan." He concluded, "There is no effective message to deter them from joining the ranks of or falling prey to the most brutal and radicalized groups, such as the Islamic State of Iraq and the Levant (ISIS/ISIL) or Jabhat al-Nusra (JN)."[54]

As such, cooperation between various intelligence and security agencies, as well as local law enforcement and community groups, will have to become closer. This relationship building entails also working with organizations and agencies which may have little to no experience dealing with security issues.[55] In essence, the foreign fighter phenomenon underscores the blurring of boundaries in contemporary national security issues. Moreover, in a borderless world, it has become evident that all forms of international issues have resonance and implications domestically. In the end, it would seem there are very few challenges 'over there' that do not have a nexus with 'over here.' Therefore, authorities and agencies at every level, i.e., municipal, provincial, and federal, both governmental and nongovernmental, must be involved and work cooperatively. As Senior Research Analyst at the NATO Council of Canada, Alexander Corbeil, concluded:

> The solutions to these issues are varied, but all include the input and collaboration between Canada's Muslim community and their leaders, private industry and government institutions beyond the narrow law enforcement focuses of CSIS and the RCMP. Those at risk of joining these groups or being influenced to carry out plots at home must be convinced of their place in Canadian society and the error of those who have already gone to fight overseas. In short, a new narrative must be created using a comprehensive approach, one which competes with the flashy propaganda machine of the [Islamic State] and other terrorist groups.[56]

As such, Professor Hegghammer argues part of any long-term policy to combat foreign fighter recruitment "must include strategies to undermine pan-Islamism, by spreading awareness of factual errors in the pan-Islamist victim narrative and by promoting state nationalisms and other local forms of identification."[57] In addition, rather than attempt to suppress the appeal of other groups or causes, a concerted effort must be made to emphasize the importance and inclusion of national civil and military institutions. The concept is to generate greater identification with the state and the individual's own society.

In this larger fight, SOF also have a potential role to play. Although most nations have strict rules with regard to the use of military forces on domestic soil, SOF are adept at working with other governmental departments and navigating the national defense and national security domains. Their precision capabilities and interagency networks and relationships make them a viable partner for dislocating, disrupting, or eliminating threats at home and abroad. For example, SOF's ability to operate in harsh and austere environments allow for SOF assistance to law enforcement agencies in rural or extreme environments. Similarly, SOF's experience in fighting networks in urban environments can also be leveraged to assist law enforcement agencies and intelligence organizations domestically.

Importantly, SOF's skills and experience can be utilized to disrupt foreign fighters, as well as any affiliated TOC organizations abroad. Either independently, or in cooperation with regional partners, SOF can use their precision kinetic and non-kinetic capabilities to counter threats at their source. By disrupting and/or eliminating training establishments, facilitating mechanisms, networks, and/or individuals at their source, SOF can substantively assist the fight against foreign fighters and potentially lone wolf terrorism (i.e., by removing the source of inspiration and radicalization of those who may be influenced by fellow nationals who have become foreign fighters or the active social media campaigns that they support).

Conclusion

Foreign fighters, the organizations and causes they support, and the lone wolf terrorism that they often inspire, represent a very real domestic and international threat. The ability of foreign fighters to travel, often relatively undetected due to their Western passports and cultural acuity, makes them a potentially hidden menace. Trained, experienced, and possibly more radicalized and/or traumatized by their experiences, they possess the potential to carry out terrorist attacks, either in groups or as a 'lone wolf' attack at home or in other Western nations. Moreover, foreign fighters and the causes they support also often act as inspiration to radicalized individuals who either through the inability to travel abroad to engage in terrorist activity, or through calculated choice, decide to conduct lone wolf attacks in their homeland.

In the end, concerted action must be taken to deter, detect, track, disrupt, and stop those who would do us harm. Notably, intelligence and law enforcement agencies operations, along with military/SOF assistance where gaps may exist, as well as increased legislation, are but some of the methodologies that can be used. Equal effort must be placed into preventing radicalization at home. In addition, effort must also be placed into disrupting and destroying extremist organizations and their leadership, which contribute to instability in the world. As President Obama explained, "Resolutions alone will not be enough. Promises on paper can't keep us safe ... Lofty rhetoric and good intentions will not stop a single terrorist attack." He concluded, "The words spoken here today must be matched and translated into action. Into deeds."[58]

Endnotes

1. J. Skidmore, "Foreign Fighter Involvement in Syria," International Institute for Counterterrorism, 2014, 9; and Center for Security Studies, "Foreign Fighters: An Overview of Responses in Eleven Countries," CSS Study, Zurich, March 2014, 4; and Jeanine de Roy van Zuijdewjn, "Fearing the Western Muslim Foreign Fighter. The connection between Fighting Defensive Jihad and Terrorist Activity in the West," MA Thesis, University of Utrecht, 2014, 59.
2. David S. Malet, "Foreign Fighters: Transnational Identity in Civil Conflicts," Ph.D. Dissertation, George Washington University, 2009, 9.
3. Barak Mendelsohn, "Foreign Fighters - Recent Trends," *Orbis*, Spring 2011, 193. Mendelsohn clarifies that a distinction must be drawn between a foreign fighter in a local conflict that is not his own country's war and "a foreign trained fighter, a local who goes to another area, receives training only, and comes back to carry out attacks elsewhere, normally in his own country."
4. Thomas Hegghammer, "The Rise of Muslim Foreign Fighters. Islam and the Globalization of Jihad," *International Security*, Vol. 35, Vol. 3 (Winter 2010/11), 57-58.
5. Ibid.
6. Skidmore, "Foreign Fighter Involvement in Syria," 9; and Center for Security Studies, "Foreign Fighters: An Overview of Responses in Eleven Countries," 4.
7. David Malet, "Why Foreign Fighters? Historical Perspectives and Solutions," *Orbis*, Winter 2010, 114. Dr. Malet noted, "It's no accident that most suicide missions in Afghanistan and Iraq were carried out by foreign fighters rather

than local militants. Fighting for what is often an abstract ideal, without having to worry about direct retaliation against their families, the Foreign Fighters need not show mercy. Some insurgent groups, such as the Islamic State of Iraq and al-Sham (ISIS) and Jabhat al-Nusra in Syria, have taken advantage of this dynamic by using foreigners to target civilians when the local combatants will not. David Malet, "In Depth: Foreign Fighters Playbook," *Foreign Affairs*, 8 April 2014, accessed 28 July 2014 at: http://www.grouph3.com/News-Analysis/in-depth-foreign-fighters-playbook.html.

8. Hegghammer, "The Rise of Muslim Foreign Fighters," 53.
9. Malet, "Why Foreign Fighters?," 97.
10. Skidmore, "Foreign Fighter Involvement in Syria," 33. The Islamic State is a Sunni Jihadist organization that was originally called the Islamic State in Iraq and the Levant or Islamic State of Iraq and Syria, depending on which version of translation is used. The confusion revolves around the interpretation of the word "Levant," which in Arabic is al-sham, which can mean the Levant, Syria or even Damascus in some circumstances. The name was changed to Islamic State in 2013.
11. Ben Winsor, "Where They Came From," *Business Insider*, 27 August 2014, accessed 29 August 2014 at: http://www.businessinsider.com/isis-is-recruiting-westerners-countries-2014-8#ixzz3Bdt8hVYI.
12. Department of Foreign Affairs, Trade and Development, Iraq Situation Report, 4 June 2015.
13. Hegghammer, "The Rise of Muslim Foreign Fighters," 53.
14. Skidmore, "Foreign Fighter Involvement in Syria," 15.
15. Thomas Hegghammer, "Should I Stay or Should I Go? Explaining Variation in Western Jihadists' Choice between Domestic and Foreign Fighting," *American Political Science Review*, February 2013, 10; and Alexander Corbeil, "Why Canada must address its foreign-fighter problem," *The Globe and Mail*, 16 July 2014, accessed 24 July 2014 at: http://www.theglobeandmail.com/globe-debate/why-canada-must-address-its-foreign-fighter-problem/article19631289/.
16. Quoted in Jez Littlewood, "Foreign Fighters: what to read to grasp the key issues," accessed 28 July 2014 at: http://npsia.wordpress.com/2014/07/03/foreign-fighters-what-to-read-to-grasp-the-key-issues/.
17. Quoted in Eric Schmitt and Thom Shanker, *Counter Strike: The Untold Story of America's Secret Campaign Against Al Qaeda* (New York: Times Book, 2011), 213.
18. Robert A. Pape, *Dying to Win: The Strategic Logic of Suicide Terrorism* (New York: Random House, 2006), 16-24.
19. Skidmore, "Foreign Fighter Involvement in Syria," 18.
20. Richard Barrett, "Foreign Fighters in Syria," The Soufan Group, June 2014, 16, 28; and Skidmore, "Foreign Fighter Involvement in Syria," 18.
21. Barrett, "Foreign Fighters in Syria," 16; and Skidmore, "Foreign Fighter Involvement in Syria," 18. A majority of fighters list their profession as student, arguably

making the case for universities having become a critical recruiting field for extremist organizations. Barrett, "Foreign Fighters in Syria," 28.

22. Hegghammer, "Should I Stay or Should I Go?," 10.
23. Steve Chase, "Al-Qaeda switching tactics, CSIS warns," 23 April 2012, *Globe and Mail.*
24. Ibid.
25. Southern Law Center, *Age of the Wolf. A Study of the Rise of Lone Wolf and Leaderless Resistance Terrorism* (Montgomery, AL: Southern Poverty Law Center, 2015), 5.
26. Quoted in Ibid., 214. A well researched 2007 New York Police Department report on radicalization described a four-phase process that transforms "unremarkable" individuals into homegrown terrorists. In the first phase, called "Preradicalization," the study notes that most homegrown terrorists have normal jobs and lead ordinary lives. The second phase, "self-identification," occurs when individuals are influenced by external or internal events and begin to explore the jihadist stream of Islam on their own, often using the Internet. The third phase, "indoctrination," transpires when an individual completely subscribes to the extremist ideology and is willing to commit violence to support the cause. The study noted that this stage is often facilitated by someone with religious influence, such as an imam or other respected figure with religious credibility, who sanction violence as religious duty. Finally, "jihadization" is attained when individuals accept it as their duty to commit violence in the name of Islam and plot to execute a terrorist act. Ibid., 213.
27. Schmitt and Shanker, *Counter Strike,* 108.
28. The term 'lone wolf' was popularized by American white supremacists Tom Metzger and Alex Curtis of the Anti-Defamation League, who believed that individual underground activity was the most effective means of advancing a cause because the 'lone wolf' leaves the least amount of collectible evidence for law enforcement, thereby limiting the likelihood of discovery or capture.
29. U.S. Government, "Rightwing Extremism: Current Economic and Political Climate Fueling Resurgence in Radicalization and Recruitment," Office of Intelligence and Analysis, IA-0257-09 (Washington: U.S. Department of Homeland Security, April 2009), 7. The autonomy of lone wolves is a key threat point. Individuals such as Theodore Kaczynski, more widely known as the "Unabomber," remained undetected for 17 years because he was able to work in isolation and take the necessary time he felt he needed.
30. Jeffery D. Simon, "What Makes Lone-Wolf Terrorists so Dangerous?", UCLA Newsroom, 18 April 2013, accessed 3 October 2014 at: http://newsroom.ucla.edu/stories/what-makes-lone-wolfe-terrorists-245316.
31. Luiza Ch. Savage, "The return of fear on the U.S.-Canada border," Macleans, 5 October 2014, accessed 5 October 2014 at: http://www.msn.com/en-ca/news/world/the-return-of-fear-on-the-us-canada-border/ar-BB7D7vU.

32. Jeremy Littlewood, "Foreign fighters: The Canadian Connection," 16 July 2014, accessed 28 July 2014 at: http://vanguardcanada.com/foreign-fighters-the-canadian-connection/.

33. BBC News, "Australia puts counter-terrorism units in airports," 27 August 2014, accessed 18 September 2014 at: http://www.bbc.com/news/world-asia-28948498. In July 2014, an 18-year-old suicide bomber from Melbourne killed several people in a market near a Baghdad mosque. In September, disturbing images of a young boy, reportedly the son of an Australian foreign fighter, holding the severed head of a Syrian soldier shocked Australia and the world. Australia has announced that it would restrict citizens from traveling to certain areas to join militant groups.

34. Schmitt and Shanker, *Counter Strike*, 112.

35. Fox News Channel, 17 June 2015.

36. Ibid., 217. On his way to the target he became spooked by surveillance by U.S. agents and he was warned by an Imam that Federal authorities had been asking questions about him. As a result, Zazi dumped his explosives and flew back home to Colorado. He was arrested several days later with the other plotters.

37. Ibid., 221. In all of these cases, terrorist groups attempted to use the foreign fighters to return home and conduct attacks because of their U.S. passports and their understanding of American culture and residency, which would allow them "to fly unimpeded below the radar."

38. The Tsarnaev brothers built their homemade pressure cooker bombs by using instructions they downloaded from an online jihadist magazine.

39. Lindsay Bever and Brian Murphy, "Gunman outside Muhammad cartoon event identified as suspected militant sympathizer," Washington Post, 4 May 2015, accessed 31 May 2015 at: http://www.washingtonpost.com/news/morning-mix/wp/2015/05/03/gunmen-shot-dead-in-texas-after-opening-fire-outside-prophet-muhammad-cartoon-show/.

40. Southern Law Center, *Age of the Wolf. A Study of the Rise of Lone Wolf and Leaderless Resistance Terrorism* (Montgomery, AL: Southern Poverty Law Center, 2015), 4, 8.

41. Romina Maurino, "Foiled terrorism plot targeted TSX, CSIS," *Globe and Mail*, accessed 6 January 2010 at: https://www.theglobeandmail.com/.

42. Michael Friscolanti, "The merciless and meticulous Toronto 18 ringleader goes to prison for life," *Macleans*, 18 January 2010, accessed 1 October 2014 at: http://www.macleans.ca/news/canada/life-sentence-for-toronto-18-ringleader/.

43. Ibid.

44. Canada, *2014 Public Report on the Terrorist Threat to Canada*, 31.

45. Canadian Press, "ISIL puts Canadians on global hit list," 22 September 2014, accessed 22 September 2014 at: http://news.ca.msn.com/canada/isil-puts-canadians-on-global-hit-list. He noted, "Like our allies we will not be cowed by threats

while innocent children, women, men and religious minorities live in fear of these terrorists."

46. Jordan Press, "Harper promises new terror measures," *The National Post*, 10 October 2014, A4.

47. Daniel Leblanc and Colin Freeze, "RCMP CSIS track foreign fighters," *The Globe and Mail*, 10 October 2014, A1. The Director of CSIS revealed, "We have no information of an imminent attack, but we have to remain vigilant, the threat is real."

48. CTV Nightly News, 20 April 2015. CSIS has noted that the phenomenon of Canadians participating in extremist activities abroad is serious. Within the last few years alone, Canadian terrorists have been implicated in attacks in Algeria, Bulgaria and Somalia, as well as Syria and Iraq. Specifically, Hassan El Hajj Hassan is wanted by Bulgarian authorities in connection with a bomb attack, on behalf of Hizballah, on a bus that killed six people and injured 35 in July 2012. In addition, two Canadian extremists participated in the January 2013 terrorist attack on the Algerian gas plant near In Amenas, Algeria; a Canadian was involved in the April 2013 attack on the Supreme Court of Somalia, which killed more than 35 people and injured dozens more; and a Calgarian, Salman Ashrafi, conducted a suicide bomb attack in an Islamic State operation in Iraq in November 2013 that killed 46 people. See Nadim Roberts, "Gregory and Collin Gordon, Calgary brothers, join ranks of Canadians fighting for ISIS," CBC News, 28 August 2014, accessed 2 September 2014 at: http://www.cbc.ca/news/gregory-and-collin-gordon-calgary-brothers-join-ranks-of-canadians-fighting-for-isis-1.2749673; and Stewart Bell, "As Many As 100 Canadians Could Be Fighting In Syria Against Assad Regime, Think Tank Says," 17 December 2013, accessed 28 July 2014 at: http://news.nationalpost.com/2013/12/17/as-many-as-100-canadians-could-be-fighting-in-syria-against-assad-regime-think-tank-says/.

49. See Canada, *Building Resilience Against Terrorism. Canada's Counter-Terrorism Strategy* (Ottawa: Public Safety Canada, 2013), 4, 13-26; Canada, *2014 Public Report on the Terrorist Threat to Canada*, 35.

50. Canada, *2014 Public Report on the Terrorist Threat to Canada*, 13.

51. MSN News, "U.N. Security Council Cracks Down on Terrorists," 24 September 2014, accessed 24 September 2014 at: http://news.ca.msn.com/top-stories/un-security-council-cracks-down-on-terrorists.

52. Center for Security Studies, "Foreign Fighters: An Overview of Responses in Eleven Countries," 6.

53. See David Malet, "In Depth: Foreign Fighters Playbook," *Foreign Affairs*, 8 April 2014, accessed 28 July 2014 at: http://www.grouph3.com/News-Analysis/in-depth-foreign-fighters-playbook.html; Schmitt and Shanker, *Counter Strike*, 213; Skidmore, "Foreign Fighter Involvement in Syria," 18; and Barrett, "Foreign Fighters in Syria," 16.

54. Pascale Siegel, "Why Should We Be Worried?," accessed 28 July 2014 at: http://www.trackingterrorism.org/article/foreign-fighters-syria-why-we-should-be-worried/why-we-should-be-worried.
55. As one example of law enforcement working with community groups, the RCMP and two Islamic groups are hoping to combat the glossy recruitment videos of extremists such as the Islamic State with a 37 page booklet aimed at youth who are in particular danger of being radicalized. The booklet, titled "United Against Terrorism," is meant to combat the idea spread by extremists that being a true Muslim means waging war on western society. It is also aimed at alerting parents to warning signs that their kids may be radicalized (i.e. sudden anti-social behavior, an interest in extremist websites and an adoption of views that are anti-women). There is also a section aimed at non-Muslims designed to educate the public at large about the difference between the Muslim religion and the violence of extremists. Steve Lambert, "Booklet helps parents recognize radical extremists," accessed 30 September 2014 at: http://preview.msn.com/en-ca/news/canada/booklet-helps-parents-recognize-radical-extremists/ar-BB6qur6.
56. Corbeil, "Why Canada must address its foreign-fighter problem."
57. Hegghammer, "The Rise of Muslim Foreign Fighters," 90.
58. MSN News, "U.N. Security Council Cracks Down on Terrorists."

Operations

Chapter 9. Borders and Security

Dr. Christian Leuprecht

Borders and Illicit Networks: Five Key Questions for Mission Success

Borders are widely taken for granted, as artifacts of state sovereignty. This chapter challenges this assumption. It makes a case for five questions every tactician, operator, and strategist should be asking about borders and networks: Do borders matter? Does the structure of cross-border networks matter? Does it matter whether the cross-border network is organized hierarchically? Does the cross-border network's purpose matter? And does the commodity being trafficked matter? Posing these questions goes some distance toward recognizing the changing role and nature of borders in globalization and how to contain their exploitation for illicit purposes. By way of example, one would expect the border to have distinct implications for gun as opposed to drug trafficking networks. Drugs tend to be illegal, and so a more complex network is required to procure, manufacture, transport them, and to bring them to a widely dispersed market of individual purchasers. Guns, by contrast, can be easier to obtain to the point where they can be a legal community. At the same time, they tend to be sold in bulk primarily for their functional purpose, rather than to maximize profit.

One of the fundamental problems those tasked with countering illicit networks face is that, while intelligence often provides a good description of

---

Dr. Christian Leuprecht is Associate Dean of the Faculty of Arts and Associate Professor in the Department of Political Science and Economics at the Royal Military College of Canada, and cross-appointed to the Department of Political Studies and the School of Policy Studies at Queen's University where he is also a fellow of the Institute of Intergovernmental Relations and the Queen's Centre for International and Defence Policy.

the flow of illicit goods, the organizational networks that facilitate these flows are poorly understood. This is not entirely surprising, since dark networks, by their very nature, are inherently difficult to study, and data collection is difficult and can, in many parts of the world, be lethal. In fact, in those places where borders and illicit networks are of greatest concern, data collection is the most difficult. So, this chapter draws instead on an emerging body of literature that strives to study borders, illicit networks, and their nexus, where data is available so as to build a body of knowledge that lends itself to application and verification elsewhere, that is, in much of the rest of the world where the phenomenon of illicit transborder networks is even more difficult to study and contain.

Border Effects

Space is a way of making sense of the world. Geographical assumptions naturalize the political segmentation of space. Borders have traditionally been understood "as constituting the physical and highly visible lines of separation between political, social and economic space."[1] But their actual significance is found in the bordering process that produces them and the institutions that manage them. These institutions "enable legitimation, signification and domination, create a system or order through which control can be exercised."[2] They politicize space and bring it under control. Quoting Painter: "The state is not only a set of institutions, but a set of understandings—stories and narratives which the state tells about itself and which make it make sense."[3] The emergence of the state has thus been contingent upon certain processes that have turned space into "state space."[4]

Border coefficients to which policy differentials across these sovereign jurisdictions give rise are considerable and their welfare implications are among the major puzzles in international economics.[5] Loesch, in *The Economics of Location*, reasoned that, according to neoclassic economics, borders created by these processes are costly because they impose barriers on free trade and the free flow of goods, labor, or skills.[6] Economic integration notwithstanding, borders continue to 'matter' because they delineate the boundaries of governments.[7] They also circumscribe social networks and human interactions.[8]

Deviant Globalization

Globalization has facilitated the emergence of transcontinental supply chains along with the expansion of illicit markets. Deviant globalization is the portion of the global economy that meets the demand for goods and services that are illegal or considered repugnant in one place by using a supply from elsewhere in the world where morals are different or law enforcement is less effective.[9] TOC has become a nefarious fixture of the global security environment. TOC's preponderance is the result of state institutions ill-suited to the challenges posed by economic and demographic inequalities, the rise of ethnic and sectarian violence, climate change, the volume of people moving across national boundaries as anything from tourists to refugees, and the growth of technology.[10]

An emerging literature posits a convergence between TOC and terrorism: terrorists are resorting to organized criminal operations to facilitate their activities, and organized crime is resorting to terrorist measures to support theirs.[11] The White House's 2011 SCTOC concludes that "criminal networks are not only expanding their operations, but they are also diversifying their activities. The result is a convergence of threats that have evolved to become more complex, volatile, and destabilizing."[12] Convergence has also improved groups' abilities to evade official countermeasures, overcome logistical challenges, identify and exploit weaknesses and opportunities in the state system, and attack that system.[13]

Terror networks are difficult to detect, and even more difficult to study accurately and comprehensively. Much of the information on individuals and their activities are either classified or unknown. Nonetheless, tracking how terrorists raise, move, store, and use money is fundamental to deter, detect, dismantle, and discourage terrorist networks. The trend has been away from studying terrorism only through the lens of individual or organizational analysis and toward the social dynamics of networks as a whole.[14] Networks make it possible for terrorist groups to overcome the inherent problems of mobilization and communication. Ergo, policymakers and security practitioners strive to know what is driving network creation, how networks operate, and how networks change over time. The analysis of longitudinal patterns of exchange between nodes is known as SNA.

Network structure may arise by design, for example, when a business constructs an organizational chart to manage coordination and governance.

However, many real-world networks are constructed because of the accumulation of pairwise connections, each of which is made locally by the two individuals concerned and sometimes with an element of serendipity. The properties of such a network are emergent, but the resulting structure is also constrained by purpose and so can be revealing of 'what works.' If the network does not contain the required actors, or if they cannot communicate as required, then the network is unlikely to be effective.

Illicit Networks

Networks are the most important unit of analysis in understanding the formation and dynamics of illicit organizations today. There has been a move away from the lens of individual or organizational analysis to an attempt to study the social dynamics of networks as a whole.[15] Social networks make it possible for illicit groups to overcome the inherent problems of mobilization and communication between large numbers of people over distances. As Thorelli states, "Organizations exist due to economies of scale and specialization, and the ability to reduce transaction costs."[16] SNA, therefore, is the study of the individual members, represented by the nodes of the network, and the relationships between these members, represented by the links. The pattern of exchanges between nodes over time is the bedrock of network analysis.[17]

As a relational approach to social interactions, SNA has emerged in the literature as an important method of analyzing and disrupting illicit networks.[18] This has not always been the case; traditionally, illicit networks were believed to be centralized and operate like hierarchical corporations.[19] By mapping out the ties between the various nodes in the group as they are, rather than how they ought to be or are expected to be, SNA theory calls this view into question.[20] Applied to various groups across different parts of the world,[21] this approach makes it possible to determine the structure and function of both the network as a whole, and the role of each person in the group in relation to others.[22]

Network Structure and Centrality

One of the most salient aspects of illicit networks that can be illustrated through the use of SNA is their structure. Network structure may arise by design, as for example, when a business constructs an organizational chart to manage coordination and governance. However, many real-world networks are constructed because of the accumulation of pairwise connections, each of which is made locally by the two individuals concerned and with an element of serendipity. The properties of such a network are emergent, but the resulting structure may also be constrained by purpose and so can be revealing of 'what works.' If the network does not contain the required actors, or if they cannot communicate as required, then the network is unlikely to be effective. Network structures matter because they dictate the flow of resources and information.

Many different types of networks—chain, hub (star), multi-player, all-channel (clique)—have been identified in the literature on illicit networks. Chain networks connect nodes in a simple path; nodes are connected only to a single neighbor in each direction, except for the initial and final nodes. Hub networks have a single central node (or perhaps a small central core of nodes) connected to other nodes in a start. The peripheral nodes have few, if any, connections to other nodes. On the one hand, the central node provides the only connection between the other nodes; therefore, it has a high level of control or leverage. On the other hand, the central node is a single point of failure, and so a vulnerability for the network. Multi-player networks feature multiple central nodes. This allows for several brokers within one network, increasing the complexity and size of multiplayer networks compared to chain or hub networks. All-channel networks are those in which most nodes are connected to most other nodes.

Hub networks contain nodes, or a small cluster of nodes, that sit at the center of three or more other nodes which themselves have few or no links. These centralized nodes are commonly referred to as hubs, and they occupy a position of influence and power because of their roles in information or materiel flow. The star network, in which a single node acts as a conduit to transmit resources and information to many other nodes, is perhaps the best-known example of a hub network. The other important structural position an actor can have in a social network is that of broker. Brokers enjoy a positional advantage within networks, as they bridge structural holes (unconnected groups of actors), and have greater access to information, opportunities, and skills. Brokers do not necessarily have to be connected to many others, as their importance derives from the fact that they bridge disconnected subparts of the network. The advantageous position of brokers in a network primarily derives from their access to diverse, non-redundant information.[23]

As Burt states, "people whose networks bridge the structural holes between groups have earlier access to a broader diversity of information and have experience in translating information across groups. This is the social capital of brokerage."[24]

Identifying hubs and brokers within a network requires the use of two key concepts: degree centrality and betweenness centrality. These two concepts, and their utility in determining prominent actors within a criminal network, were outlined by Morselli.[25] Degree centrality is the simple measurement of the volume of contacts each node possessed within a network. It is a measure of the number of an actor's immediate neighbors. Betweenness centrality is a measure of an actor's position within a network. Actors with a high betweenness measure frequently sit on the shortest path between two other actors.[26] This measure captures the extent to which certain nodes connect parts of the network that would otherwise have poorer contact, perhaps even no contact. These important individuals are brokers insofar as they control and mediate the flow of information and resources between unrelated parts of a network.[27] Specific characteristics of an individual's role in a network are associated with specific centrality measures, as depicted in the matrix in Table 1.

High degree centrality and low betweenness centrality *Well-informed member of a subgroup*	High degree centrality and high betweenness centrality *Subgroup leader*
Low in both degree and betweenness centrality *Foot soldier*	Low degree centrality and high betweenness centrality *Ideal broker*

Table 1. Centrality Matrix[28]

An individual with both high degree centrality and high betweenness centrality has the advantages of being able to control the flow of information between different subnetworks, and being able to reach a large number of individuals quickly. In social networks, the degree achievable by such a subgroup leader is limited by cognitive and span of control issues—in criminal and illicit networks, the individuals in these positions are aware of the vulnerability their high degree creates for disruption by their arrest or removal. Thus there are pressures to limit the degree of any one individual. On the other hand, for criminal and illicit social networks, lack of trust creates a countervailing pressure toward increased degree for network leaders to reduce the threats of surprise leadership challenges or group splintering. In such an organization, therefore, an individual with both high degree centrality and high betweenness centrality is likely to be the leader, at least of a major subgroup.

However, Morselli's study of the Hell's Angels biker gang in the province of Quebec, Canada, discovered that the most advantageous position in a network is that of the 'ideal broker,' an individual with low degree centrality but high betweenness centrality measure. Such a broker is able to take strategic actions within the network, and to provide access to information and resources that others cannot reach, but does not necessarily have high volumes of communication or management with which to contend. Also, such individuals' lower degrees make them less visible.[29]

Resilience and Hierarchy

Networks provide resiliency for criminal operations. Duijn, Kahirin, and Sloot found that criminal networks can become even stronger after being targeted by law enforcement, meaning police action against criminal networks must be undertaken carefully.[30] There is a consistent tradeoff between security and efficiency in dark networks, and criminal organizations often opt for efficiency, given the limited time-to-task under which they operate.[31] While one often thinks of criminal networks as being rigid in structure, the extent to which they adapt is key to their survival.[32] In response to pressure from law enforcement, networks have been found to decentralize their structure, though this can often still leave them vulnerable.[33] Indeed, Bright and Delaney found that over time the centralities of nodes in a drug trafficking network changed as individuals changed roles to meet the evolving needs of the network.[34] As a result, networks are often much more decentralized than we might assume. Kenney found that the Colombian drug trade was based on a series of small fluid networks, rather than monolithic cartels.[35] The resilience of drug trafficking networks, namely their ability to adapt, can make them a challenging opponent for law enforcement. Indeed, hierarchy runs counter to the very characteristics of networks, which are heralded as "temporary, dynamic, emergent, adaptive, entrepreneurial and flexible structures," a "cutting-edge design."[36] Similarly, the rigid depiction of networks contrasts starkly with networks as an organizational structure that consists of "operatives [who] are highly adaptive, compartmentalised [and] mobile."[37]

Functional Differentiation

The nature and score of centrality measures is related to an illicit network's function—the structure of the network reflects the function that it serves.[38] Comparing al-Shabaab networks, fundraising networks exhibited a hub-type structure, whereas recruitment networks were of an all-channel type where most nodes are connected to most other nodes. Moreover, fundraising networks were transcontinental, that is, they crossed international borders. Therefore, fundraising networks are not impeded by the physical proximity of individuals in the network to one another but rely heavily on the actions of brokers.[39]

Commodities

In a study of gun-trafficking across the Canada-U.S. border, I have shown how commodity affects network structure; cross-border gun trafficking networks take the form of simple chain networks or slightly more advanced hub networks.[40] Given the availability of legal guns in the United States, it is understandable why chain networks are so prevalent; cross-border trafficking is as simple as crossing the border. Chain networks also appear to be easy to disrupt. Simply removing one actor breaks the chain. Actors in chain networks are relatively equal in terms of centrality. This results in an equal flow of information and resources through the sequential actors in a chain network; consequently, when one actor is compromised, so are the others. As a result, few members of chain cross-border gun trafficking networks escape arrest when their ring is discovered and disbanded by law enforcement.

By contrast, hub networks have greater capacity. Mules tend to be low in both degree and betweenness centralities, coordinated by a broker who bridges the gap between the supply of guns being delivered by the mules and the buyer who often takes the form of an organized crime syndicate. The bridge provided by these brokers establishes them as the crux of the network by virtue of their high betweenness centrality. Without their presence, the supply of guns can be easily cut off. The various gun trafficking networks observed met their ends when brokers were compromised and arrested. While targeting brokers appears to be an effective way of disrupting the cross-border gun trafficking networks that take the form of hub networks, the more challenging task is to ensure that other brokers do not take their place and networks spring up to fill the void.

In the same article I also hypothesize that transborder gun trafficking networks take simpler forms than transborder drug trafficking networks, as manifested in different network structure. For both kinds of networks, the actual cross-border piece is analogous: a number of parallel border-crossers and a single collector node on the other side. That there is only one collector node is likely a function of trust; if mules had more than one person to whom to report, it would become too easy for them to cross the border and go into business for themselves. It is surprising that collectors do not set up cutouts between themselves and the arriving mules. That would make them much harder to detect after interception.

There are two differences between the two kinds of networks. The first arises from the ease of obtaining the commodity. Since guns are readily available, no structure is needed to handle acquisition. In contrast, obtaining drugs requires access to a global pipeline, thus a much more sophisticated and extensive acquisition process, and a network to support it. The second difference is on the downstream side and derives from the quantum of the objects being transported. On the one hand, profit per gun is large, and collectors do not bother to establish distribution networks. This is their Achilles heel: they get ratted out by those to whom they sell. On the other hand, drugs are sold in much smaller quantities than guns, forcing collectors to build and work through distribution networks, which protect them from being ratted out because they are further from the people who get arrested.

Project Corral, a large anti-gang operation launched in 2009 by the Toronto police with the support of over 1,000 police officers of the police services of Ontario, illustrates the efficacy of network analysis to bring down transnational crime. Local gangs such as the Falstaff Crips and the Five Point Generals were closely linked to the Jamaican criminal organization known as the Shower Posse, which was supplying drugs to the gangs. Police in 19 jurisdictions seized cash, arms, drugs, and other contraband.[41]

In Project Corral, drug trafficking took the form of a complicated multiplayer network, with numerous influential actors. The gun trafficking network that was being operated by the same organization was different. Despite the capacity that the Shower Posse had to run a complex network, their transborder gun trafficking network was unsophisticated. The network hinged on only one or two key brokers to maintain a steady flow of weapons across the border.

Just as Morselli and Calderoni had found, the Shower Posse leadership seems to have reduced their degree centrality and likelihood of arrest by avoiding participation in street level crime.[42] Much like Bright and Delaney had found, members of the Shower Posse changed their role over time as the gang adapted to new circumstances.[43] This flexibility that has been studied in criminal networks in the past helps generate the hypothesis that the Shower Posse adopted a more complex network in response to the presence of borders.

As the Shower Posse grew outward from Jamaica, greater complexity was required in the structure of the network. As the scope of the operation grew, it was no longer possible to run a hub network based solely out of Tivoli Gardens in Kingston, Jamaica. Vivian Blake became a broker in his own right coordinating the New York chapter of the Shower Posse, and this model was repeated in the various countries into which the gang expanded. Moving illicit commodities across borders required the Shower Posse to have a presence on both sides. While the Shower Posse may have acted as a wholesaler of drugs to local street gangs, they only did so once they themselves had transported the drugs over the border to their local presence. Creating such a complex network was aided by the Jamaican diaspora in the United States and Canada, which eased the process of creating local Shower Posse chapters. The Shower Posse also took advantage of political corruption in Jamaica to ensure security for their network, allowing for a greater focus on the coordination necessary to traffic such a large quantity of illicit commodities across multiple borders.

With respect to gun trafficking networks, the transaction costs imposed by the border appear low compared to the vast markets of opportunity the border creates. The prerequisites are a ready supply of guns on one side of the border and someone who is willing to purchase guns and then bring them across the border. Single cases of gun trafficking, or simply individuals who do not require a larger network to profit from gun trafficking, are thus quite possible.

The ease with which individuals can cross and the large supply of legal guns in the United States seems to allow for the proliferation of many small, unsophisticated gun trafficking networks. That explains why even the gun trafficking networks with significant organized crime connections do not appear to differ substantially from those operated by a handful of individuals.

Conclusion

The Jamaican Shower Posse is emblematic of just how difficult it is to contain TOC networks. The sophistication of the Shower Posse's operations, combined with its political influence in its home country, made it tedious for authorities to eradicate. When the power of the state is weak, gangs can fill the void by performing state-like functions, which endears them to locals. Attempting to bring down such organizations through the prosecution of low- and mid-level dealers is ineffective, as the networks are too complex to be disrupted by the loss of one or two nodes with limited connections. How was the Shower Posse's kingpin Christopher "Dudus" Coke able to live out in the open and avoid capture for so long? The complexity of the Shower Posse's all-channel network shielded him from prosecution. And how did the Jamaican Shower Posse that Coke headed extend its reach all the way from Jamaica to the U.S. and Canada? Through a network that had to have sufficient players to procure ingredients, manufacture drugs, traffic them, and bring them to market in an environment where the commodity is illicit yet requires many small-scale buyers.

The nature of the network was a function of the borders the Shower Posse needed to cross in its activities. If network science can prove that transnational drug trafficking operations reinforce their resilience by making the structure of their networks more complex in response to the challenge posed by trafficking illicit commodities across national boundaries, then these networks can be better understood and targeted. Such a complex network was a response to the borders the Shower Posse needed to cross in its activities.

The Shower Posse used a complex multiplayer network backed by political connections in Jamaica to create a robust criminal enterprise. Criminal organizations like the Shower Posse are impervious to being targeted through street-level dealers. The complexity of the network means that it will continue to operate with little disruption, as those in charge and other dealers, mules, and distributors are still left intact. Nor does such an approach address the strong political connections that shield powerful gangs. To target criminal organizations that operate on the international scale of the Shower Posse effectively, gang leadership—those with high degree centrality but low between centrality—must be identified and targeted to disrupt the network and bring an end to its activity. Since clientèlism is corrosive to democracy, measures to curb political corruption that allows such gangs to flourish in the host countries in which they are based will pay dividends by making organized crime networks less resilient.

The structure of networks matters to deterring, detecting, and disrupting them. Transnational networks are already known to pose a unique challenge due to the fact that they occupy multiple jurisdictions. If they have also adapted to the challenge posed by straddling borders by adopting more complex network structures, then the threat they pose to law enforcement is even greater. Removing only a few actors from the network will do little to compromise it, especially if they are low- or mid-level operatives. Rather, concentrated efforts to remove the gang's leadership across the network is necessary for its operation to be disrupted.

If the border imposed high marginal costs on trafficked goods, we would expect to see complex networks. This appears to be the case for drugs, but less so for guns. Trafficking in drugs requires volume to turn a profit; trafficking in guns does not. Ergo, policy differentials across borders and the markets of opportunity they create matter. A commodity that is legal on one side of the border but not the other is subject to trafficking for direct or indirect gain by means of relatively simple chain or hub networks. Complex multi-player networks appear necessary, by contrast, when a good is illegal on either side of the border and profit is a function of volume.

Mapping the structure of illicit networks is imperative to understanding how best to target and disrupt these networks. It is simple to compromise the actors in a chain network, and if a broker can then be located in a hub network, he or she can be targeted, and the whole network will be disrupted. This is especially true of brokers who traffic guns since they have a high degree centrality to match their high betweenness centrality; that is, while they bridge important structural gaps in networks, they are widely known throughout the network, as they are the key contact for the other actors. For instance, few gun traffickers are ideal brokers precisely because they are subject to being identified by so many other members of the network, which makes them an easy target for identification by law enforcement. The unsophisticated nature of these networks also helps to explain why they are plentiful. Provided that one knows how to tap into the market, they are simple to set up and simple to operate. For this reason, the real challenge of understanding cross-border trafficking networks is not how to target brokers and the networks they connect, but how to discourage people from becoming brokers and enabling networks to regenerate. For this reason, the intelligence-led policing model focuses on disrupting and dismantling networks by concentrating scarce resources on brokers in the form of dynamic network analysis and target selection.

In conclusion, borders, networks, and their nexus matter to detecting, dismantling, and deterring organized illicit activity. Borders impose countervailing transaction costs but also open up markets of opportunity; dynamics which affect the formation and structure of transborder networks. While assumptions about hierarchical organizational structures abound, empirical evidence suggests that structures are both flatter and more autonomous than generally assumed. A network's function has a significant bearing on its structure; therefore, an understanding of its structure can provide insights into its purpose, and vice versa. Finally, the nature of the commodities that flow through a network affect both the structure of the network, as well as the way networks diffuse across boundaries. A more aggressive research agenda on border and networks should allow us to answer some of these questions with greater precision. Carefully analyzing the connection between borders and networks may be a simple, yet important contribution to mission success for those tasked with containing the genesis and diffusion of illicit transborder networks and their second-order effects.

Endnotes

1. David Newman, "The Lines That Continue to Separate Us: Borders in Our 'Borderless' World," *Progress in Human Geography*, 30 (2), 2006, 144.
2. Ibid., 149.
3. Joe Painter, *Politics, Geography and "Political Geography"* (London: Arnold, 1995), 47.
4. Neil Brenner, Bob Jessop, Martin Jones, and Gordon Macleod, *State/Space: A Reader* (Oxford: Blackwell Publishing, 2003).
5. Maurice Obstfeld and Kenneth Rogoff, "The Six Major Puzzles in International Macroeconomics: Is There a Common Cause?" in *NBER Macroeconomics Annual*, 2000, Vol. 15, B.S. Bernanke and K. Rogoff, eds. (Cambridge, MA: MIT Press, 2001), 339-412.
6. August Loesch, *The Economics of Location*. 2nd rev. ed. (New Haven: Yale University Press, 1954).
7. John F. Helliwell, *How Much Do National Borders Matter?* (Washington, D.C.: Brookings Institution Press, 1998), and *Globalization and Well-Being* (Vancouver: UBC Press, 2002).
8. Geoffrey Hale and Monica Gattinger, *Borders and Bridges: Canada's Policy Relations in North America* (Don Mills, ON: Oxford University Press, 2010).

9. Nils Gilman, Jesse Goldhammer and Steve Weber, *Deviant Globalization: Black Market Economy in the 21st Century* (New York: Bloomsbury Academic, 2011); and Moises Naim, *Illicit: How Smugglers, Traffickers, and Copycats are Hijacking the Global Economy* (New York: Knopf Doubleday, 2006).
10. Louise I. Shelley, *Dirty Entanglements: Corruption, Crime, and Terrorism* (Cambridge: Cambridge University Press, 2014).
11. Michael Miklaucic and Jacqueline Brewer, *Convergence: Illicit Networks and National Security in the Age of Globalization* (Washington, D.C.: National Defence University Press, 2013); Colin P. Clarke, *The Financing of Terrorism, Insurgency, and Irregular Warfare* (Santa Barbara: Praeger Security International, 2015), 70-94; Lyubov Mincheva and Ted R. Gurr, *Crime-Terror Alliances and the State. Ethnonationalist and Islamic Challenges to Regional Security* (Abingdon: Routledge, 2012); Phil Williams, "Terrorist financing and organized crime: nexus, appropriation, or transformation?" In Thomas J. Biersteker and Sue E. Eckert (eds.), *Countering the Financing of Terrorism* (New York: Routledge, 2008), 126-149; Gregory F. Treverton, Carl Matthies, Karla J. Cunningham, Jeremiah Goulka, Greg Ridgeway and Anny Wang, *Film Piracy, Organized Crime, and Terrorism* (Santa Monica, CA: RAND, 2009); Peter Alldridge, "Acquisition and Deployment of Money for Terrorism," in *Money Launder Law: Forfeiture, Confiscation, Civil Recovery, Criminal Laundering and Taxation of the Proceeds of Crime* (Oxford and Portland: Hart Publishing, 2003), 215-222.
12. The White House, *Strategy to Combat Transnational Organized Crime: Addressing Converging Threats to National Security* (Washington, D.C.: The White House, July 2011).
13. Nick McKenzie and Richard Baker, "Terrorists Taking Cut Of Millions In Drug Money," *The Age*, 23 January 2014.
14. A. Perliger and A. Pedahzur, "Social Network Analysis in the study of terrorism and political science," *PS: Political Science and Politics* 44(1) 2011, 45-50.
15. Ibid., 45.
16. Hans B. Thorelli, "Networks: Between Markets and Hierarchies," *Strategic Management Journal 7*, no. 1, 1986, 37.
17. Walter W. Powell, "Neither Market nor Hierarchy: Network forms of Organization," *Research in Organizational Behavior,* vol. 12, 1990, 295.
18. For example, see Marc Sageman, *Understanding Terror Networks* (Philadelphia: University of Pennsylvania Press, 2004); Sean Everton, *Disrupting Dark Networks* (Cambridge, Cambridge University Press, 2013); Renée van der Hulst, "Terrorist networks: The threat of connectivity," in John Scott and Peter J. Carrington (eds), *The SAGE Handbook of Social Network Analysis* (London: SAGE, 2011), 256-270.
19. Cynthia Stohl and Michael Stohl, "Networks of Terror: Theoretical Assumptions and Pragmatic Consequences," *Communication Theory* 7, 2007, 95.
20. Roberta Belli, Joshua D. Freilich, Steven M. Chermak, and Katharine Boyd, *Exploring the U.S. Crime-Terror Nexus: Terrorist Networks and Trade Diversion,*

National Consortium for the Study of Terrorism and Responses to Terrorism Research Brief October (College Park, MD: University of Maryland, 2014).

21. Luke M. Gerdes, Kristine Ringler and Barbara Autin, "Assessing the Abu Sayyaf Group's Strategic and Learning Capacities," *Studies in Conflict & Terrorism* 37, no. 3, 2014, 267-293; Olivier Walther, Dimitris Christopoulos, "Islamic terrorism and the Malian rebellion," *Terrorism and Political Violence*, 2014, 1-23; Kathleen M. Carley, Ju-Sung Lee and David Krackhardt, "Destabilizing Networks" *Connections* 24, no. 3, 2002, 79–92; Stuart Koschade, "A Social Network Analysis of Jemaah Islamiyah: The Applications to Counter-Terrorism and Intelligence" *Studies in Conflict and Terrorism* 29, no. 6, 2006, 559–575; Ian A. McCulloh, Kathleen M. Carley and Matthew Webb, "Social Network Monitoring of Al-Qaeda," *Network Science* 1, no. 1, 2007, 25–30; and Perliger and Pedahzur, 45–50.
22. Perliger and Pedahzur, 46.
23. Ronald S. Burt, *Structural Holes. The Social Structure of Competition* (Cambridge MA: Harvard University Press, 1992); Ronald S. Burt, *Brokerage and Closure. An Introduction to Social Capital* (Oxford, Oxford University Press, 2005); Spiro Emma, Acton Ryan, Butts Carter, "Extended structures of mediation: Re-examining brokerage in dynamic networks," *Social Networks* 35, 2013, 130–143.
24. Ronald S. Burt, "Structural Holes and Good Ideas," *The American Journal of Sociology* 110, no. 2, 2004, 354.
25. Carlo Morselli, *Inside Criminal Networks* (New York: Springer, 2009).
26. Carlo Morselli, "Assessing Vulnerable and Strategic Positions in a Criminal Network," *Journal of Contemporary Criminal Justice* 26, 2010, 382-384; Christina Prell, *Social Network Analysis: History, Theory & Methodology* (London: Sage, 2012), 97-107.
27. Morselli, "Assessing Vulnerable and Strategic Positions in a Criminal Network," 382-384.
28. This matrix of Morselli's centrality measures first appeared in Christian Leuprecht and Kenneth Hall, "Networks as Strategic Repertoires: Functional Differences among Al-Shabaab terror cells," *Global Crime*, 2013.
29. Ibid.
30. P. Duijn, V. Kashirin, and P. Sloot, "The Relative Ineffectiveness of Criminal Network Disruption," *Scientific Reports*, 2014, 4.
31. C. Morselli and K. Petit, "Law-Enforcement Disruption of a Drug Importation Network," *Global Crime* 8(2), 2007, 109-130.
32. C. Morselli and J. Roy, "Brokerage Qualifications in Ringing Operations," *Criminology* 46(1), 2008, 71-98.
33. Morselli and Petit, "Law-Enforcement Disruption of a Drug Importation Network."
34. D. Bright and J. Delaney, "Evolution of a Drug Trafficking Network: Mapping Changes in Network Structure and Function Across Time," *Global Crime* 14(2), 2013, 238-260.

35. M. Kenney, "The Architecture of Drug Trafficking: Network Forms of Organization in the Colombian Cocaine Trade," *Global Crime* 8(3), 2007, 233-259.

36. S.R. Corman, "Using activity focus networks to pressure terrorist organizations," *Computational & Mathematical Organization Theory* 12, 2006, 35.

37. M. Stohl, "Networks, terrorists, and criminals: the implications for community policing," *Crime, Law and Social Change*, 50, 2008, 66.

38. C. Leuprecht and K. Hall, "Networks as strategic repertoires: Functional differentiation among Al-Shabaab terror cells," *Global Crime* 14(2-3), 2013, 287-310. See also: Anthony Masys, "Networks and Network Analysis for Defence and Security," *Lecture notes in Social Networks* series, 83-120.

39. Christian Leuprecht and Kenneth Hall, "Why Terror Networks are Dissimilar: How Structure Relates to Function," in A.J. Masys (ed.), *Networks and Network Analysis for Defence and Security* (Switzerland: Springer, 2014), 93-94; see also Leuprecht and Hall, "Networks as Strategic Repertoires."

40. C. Leuprecht and A. Aulthouse, "Guns for Hire: Mapping Canada-US cross-border gun trafficking networks," *Journal of Criminology, Criminal Justice, Law and Society* 15(3), December 2014, 57-74.

41. Mary Vallis, "Massive police raids executed across Ontario: Project Corral," *National Post*, 4 May 2010, accessed 6 August 2015 at: http://news.nationalpost.com/news/massive-police-raids-executed-across-toronto-ottawa-this-morning.

42. F. Calderoni, "The Structure of Drug Trafficking Mafias: The 'Ndrangheta and Cocaine," *Crime, Law & Social Change* 58(3), 2012, 321-349; and C. Morselli, "Assessing vulnerable and strategic positions in a criminal network," *Journal of Contemporary Criminal Justice*, 26(4), 2012, 382– 392.

43. Bright and Delaney, "Evolution of a Drug Trafficking Network."

Chapter 10. An Ontological Framework for Understanding the Terror-Crime Nexus

Colonel Bill Mandrick, Ph.D.

Introduction

> Transnational terrorists and criminals may collaborate, appropriate shared tactics, and otherwise benefit from interaction, resulting in bolstered capabilities, enhanced organizational infrastructure, improved access to resources, and expanded geographic reach. Historical examples also indicate that terrorist and transnational criminal groups may evolve, converge, transform, or otherwise alter their ideological motivations and organizational composition to appear similar.[1]

In an increasingly interconnected world, people within one complex network can seek out and merge with people in other complex networks, thereby creating a network of networks. The merging of complex social networks (i.e., globalization) can contribute to the betterment of society by increasing capabilities to solve difficult problems. However, the merging of complex networks also results in deviant globalization where illicit "cross-border economic networks produce, move, and consume things as various as narcotics and rare wildlife, looted antiquities and counterfeit goods, dirty money and toxic waste, as well as human beings in search of undocumented work and unorthodox sexual activities."[2] Terrorist and insurgent networks benefit from deviant globalization by connecting with TCOs in order to employ their capabilities and generate funds to continue operations. The National Security Council's SCTOC states that:

Colonel Bill Mandrick, U.S. Army Reserve, Ph.D., is a civil affairs officer who has planned and led civil-military operations in austere environments. He is a Senior Fellow at the Center for Special Operations Studies and Research within the Joint Special Operations University. His current research pertains to the use of property information for joint intelligence processes.

> Terrorists and insurgents increasingly are turning to TOC to generate funding and acquire logistical support to carry out their violent acts. The Department of Justice reports that 29 of the 63 organizations on its FY 2010 Consolidated Priority Organization Targets list, which includes the most significant international drug trafficking organizations (DTOs) threatening the United States, were associated with terrorist groups. Involvement in the drug trade by the Taliban and the Revolutionary Armed Forces of Colombia (FARC) is critical to the ability of these groups to fund terrorist activity. We are concerned about Hizballah's drug and criminal activities, as well as indications of links between al-Qa`ida in the Lands of the Islamic Maghreb and the drug trade. Further, the terrorist organization al-Shabaab has engaged in criminal activities such as kidnapping for ransom and extortion, and may derive limited fees from extortion or protection of pirates to generate funding for its operations.[3]

The purpose of this chapter is to provide a semantic framework (i.e. ontology) that will enhance the counterterrorism analyst's ability to understand and describe the nexus between transnational terrorist organizations and criminal organizations. This framework will enhance the management of disparate data as it becomes available to the analyst, planner, or decision maker in CT operations.[4] For this chapter, consider a case where a CT analyst has access to already existing data sets on terrorist activities and related events. These disparate data sets are structured, meaning that they are created in compliance with a predefined data model, for use in some specific information system. The result is that each data set is 'stove-piped' or incompatible with the other data sets. The analyst also has access to near-real-time tactical level reports from various sources, ranging from maneuver unit situation reports to site exploitation team forensics reports. Much of this data is unstructured—i.e., it is in report format consisting of text strings, descriptive sentences, paragraphs, and images. With some effort, this data can be 'tagged' with category terms taken directly from the ontological framework, making it interoperable with any other data that is similarly aligned.

The enhancement of data in this way, (a) generates alignment because it adds consistent descriptions to the inconsistent sets of codes, and (b) does this in a way that allows software tools to reason over the result. Enhanced data can be ingested into an IT system that facilitates analysis based upon a powerful set of algorithms. With properly collated data, the system can graphically depict correlations between physical objects, realizable attributes, locations, and events. The ultimate purpose of the system is predictive analytics, based upon large amounts of aligned data. The CT analyst can use the following ontological framework to organize the disparate data so that it can be analyzed to identify these hidden correlations, and then disseminated (shared) across the larger CT enterprise.

Ontology is the disciplined study of what is, of the kinds and structures of objects, properties, events, processes, and relations in every area of reality. From this philosophical perspective, ontology seeks to provide a definitive and exhaustive classification of entities in all spheres or domains of being. As a theoretical discipline concerned with accurately describing the taxonomy of all things that exist, philosophical ontology is synonymous with classical metaphysics. High quality ontologies define, disambiguate, and relate the objects, attributes, and processes that make up any domain of interest.

The ontology to be described here is intended to enhance intelligence processes, and assist analysts, planners, and decision makers in analyzing, defining, and representing the elements (nodes and links) that make up any dynamic human network, such as an insurgent network, terrorist network, or TOC network. The intelligence products produced by using this framework will inform the find, fix, finish, exploit, analyze, and disseminate process as prescribed in Joint Publication 3-26, *Counterterrorism*.

The proposed semantic framework provides an extendable vocabulary to which any data source (information element) that describes some aspect of the complex operational environment (OE) could be translated. In order to provide coverage of any aspect of an OE, the chosen strategy was to build a framework sufficient to express general-level facts but which were easily extended in a consistent and uniform way to ontologies having finer grained content as needed.[5] Ultimately, the framework is designed to assist analysts and planners in understanding the form of these networks, as well as their capabilities and functions, and the types of activities they participate in. These elements form nodes and links (relationships) in a graph-like structure, which can be used for enhanced reasoning by both humans and machines.

Information that is managed in accordance with this semantic framework will be in a format that will assist operations planners in their reasoning and decision making about clandestine human networks within the complex OE. The proposed framework consists of a small set of ontologies, which are more effective for organizing information because they consist of more specific categories than those that make up overly general acronyms or mnemonic memory devices. For example, the civil information management process contributes to understanding the OE through civil considerations analysis, which focuses upon areas, structures, capabilities, organizations, people—the ASCOPE mnemonic. Systems analysis identifies centers of gravity (e.g., some highly connected individual) and enhances situational understanding by focusing upon political, military, economic, social, information, infrastructure (PMESII), systems within the OE, as well as the physical environment and time.[6] These two mnemonic devices can be combined into the following crosswalk matrix, which helps analysts and operators to organize information about the OE.

Nodal analysis reveals the interrelationship between people, organizations, entities, and locations. The individual nodes represent complex relationships between a person, place, or physical thing that are a fundamental component of a system and link the behavioral, physical, or functional relationships between the nodes. Critical nodes are those identified as being essential and whose disruption or removal becomes a single trend analysis point failure. Link analysis is the process of identifying and analyzing relationships between personnel, contacts, associations, events, activities, organizations, and networks to determine key or significant links. Analysts

	Political	Military	Economic	Social	Infrastructure	Information
Areas	Political Areas	Military Areas	Economic Areas	Social Areas	Infrastructure Areas	Information Areas
Structures	Political Structures	Military Structures	Economic Structures	Social Structures	Infrastructure Structures	Information Structures
Capabilities	Political Capabilities	Military Capabilities	Economic Capabilities	Social Capabilities	Infrastructure Capabilities	Information Capabilities
Organizations	Political Organizations	Military Organizations	Economic Organizations	Social Organizations	Infrastructure Organizations	Information Organizations
People	Political People	Military People	Economic People	Social People	Infrastructure People	Information People
Events	Political Events	Military Events	Economic Events	Social Events	Infrastructure Events	Information Events

Table 1. The PMESII—ASCOPE crosswalk matrix assists in the identification of types of data needed for civil considerations and systems analysis. This same data will become inputs for nodal and link (relationship) analysis. Source: Author

use link analysis to determine who is involved with whom and how they are involved.[7]

The ontological framework described here is for the enhancement of civil information management, especially where it is concerned with nodal and link analysis for complex social systems. Applied ontology offers a strategy for the organization of scientific information in computer-tractable form, drawing on concepts not only from computer and information science but also from linguistics, logic, and philosophy.[8] Data sets that are aligned with this ontological framework can be integrated with each other so that they become one larger enhanced data set. Enhanced data sets facilitate analytics intended to show hidden correlations between system variables. As will be described later in the chapter, data that is aligned in this way (i.e. enhanced data) contributes to JIPOE.

Definitions for Terrorism and Transnational Crime

Terrorism is defined in Joint Publication (JP) 3-07.2, *Antiterrorism*, as:

> the unlawful use of violence or threat of violence to instill fear and coerce governments or societies. Terrorism is often motivated by religious, political, or other ideological beliefs and committed in the pursuit of goals that are usually political.[9]

The ontology of terrorism starts by decomposing this definition into discrete elements such as an act of violence, the belief that is a motive for that action, the government or group of people that are the target of the action, and the goal of the action. What is implied in this definition is that there are relationships between individual persons, and that these relationships make up a terrorist organization that is a participant in the act of terrorism. Furthermore, the terrorist organization is the bearer of certain dispositions and capabilities—see Figure 1.

TOC is defined as "those self-perpetuating associations of individuals who operate transnationally for the purpose of obtaining power, influence, monetary and/or commercial gains, wholly or in part by illegal means, while protecting their activities through a pattern of corruption and/or violence, or while protecting their illegal activities through a transnational organizational structure and the exploitation of transnational commerce or communication mechanisms. There is no single structure under which transnational organized criminals operate; they vary from hierarchies to clans, networks, and cells, and may evolve to other structures. The crimes they commit also vary …"[10] This definition includes social relationships between individual persons in an organization or structure, such as: associate of, subordinate of, relative of, commander of, et cetera. The definition also describes the actions, motivations, and locations of these organizations. TOC organizations will also have certain realizable attributes such as dispositions to act and enabling capabilities. Dispositions and capabilities are realized by certain activities.

The basic ontological structure of any organization has similar elements—see figure 1. The organization will be the bearer of certain realizable attributes. Any realizable attribute is realized by the organization being a participant in some action (or set of actions), which has location in some geopolitical territory. The relationships depicted in italics, are a nexus, bond, link, or junction; a means of connection between things or parts.[11]

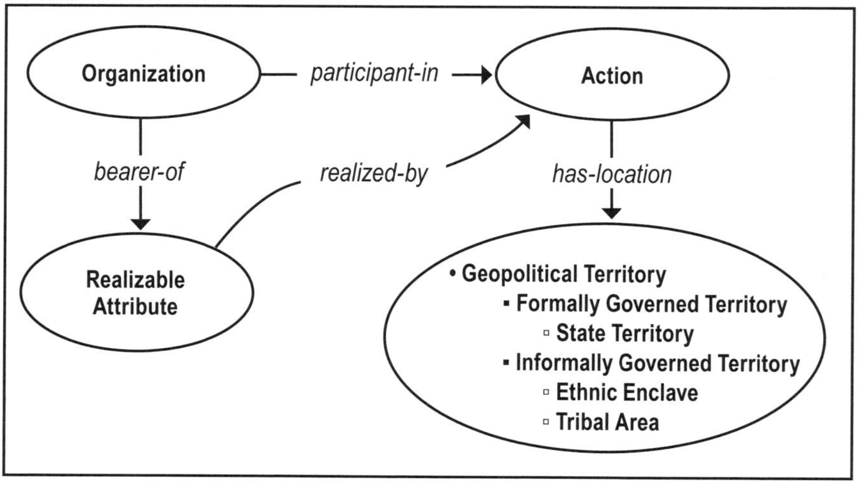

Figure 1. The basic ontology of an organization. Source: Author

An Ontological Framework for Understanding the Transnational Terrorism-Crime Nexus

> At issue is how to make sense of this complex crime-terrorism phenomenon and oversee the implementation of cross-cutting activities that span geographic regions, functional disciplines, and a multitude of policy tools that are largely dependent on effective interagency coordination and international cooperation.[12]

> The wrong ontology and epistemology, largely based on mirror-imaging information-age network theories onto clandestine cellular networks, have led many network and counter-network theorists astray. Most theorists and practitioners cognitively mirror information-age networks to clandestine cellular networks, which is largely incorrect. Failure to understand the aspects of clandestine cellular networks has huge implications to both the way network theorists study and model networks, as well as how network attack theorists recommend defeating clandestine cellular networks.[13]

CT analysts, operations planners, and command staff members are charged with making sense of the complex OE, which is doctrinally defined as a composite of the conditions, circumstances, and influences that affect the employment of capabilities and bear on the decisions of the commander.[14] In order to analyze and define the elements of an OE, analysts adopt a systems perspective that is intended to provide an understanding of significant relationships within interrelated PMESII, and other systems relevant to a specific joint operation.[15] The systems perspective facilitates the identification of nodes and links within and between systems. These nodes and links (relationships) are the subjects of planning and decision making processes—e.g., observing, targeting, engaging, and influencing.

For the purposes of analyzing the nexus between TCOs and terrorist organizations, the analyst must adopt the same systems perspective toward those types of systems, and then identify the nodes and links that connect them to other PMESII systems—e.g., some person involved in TOC may be associated with some political organization. This section describes a semantic framework (i.e., ontology) that is intended to assist analysts, planners, and decision makers in analyzing, defining, and representing the elements (nodes and links) that make up an OE, where TCOs and terrorist organizations sometimes collaborate in order to advance their interests through criminal activities.

The proposed framework consists of a small set of ontologies for organizing information in conjunction with the PMESII and ASCOPE crosswalk matrix. It provides an extendable vocabulary to which any data source (information element) that describes some aspect of the OE could be translated. In order to provide coverage of any aspect of an OE, the chosen strategy was to build a framework sufficient to express general-level facts but which were

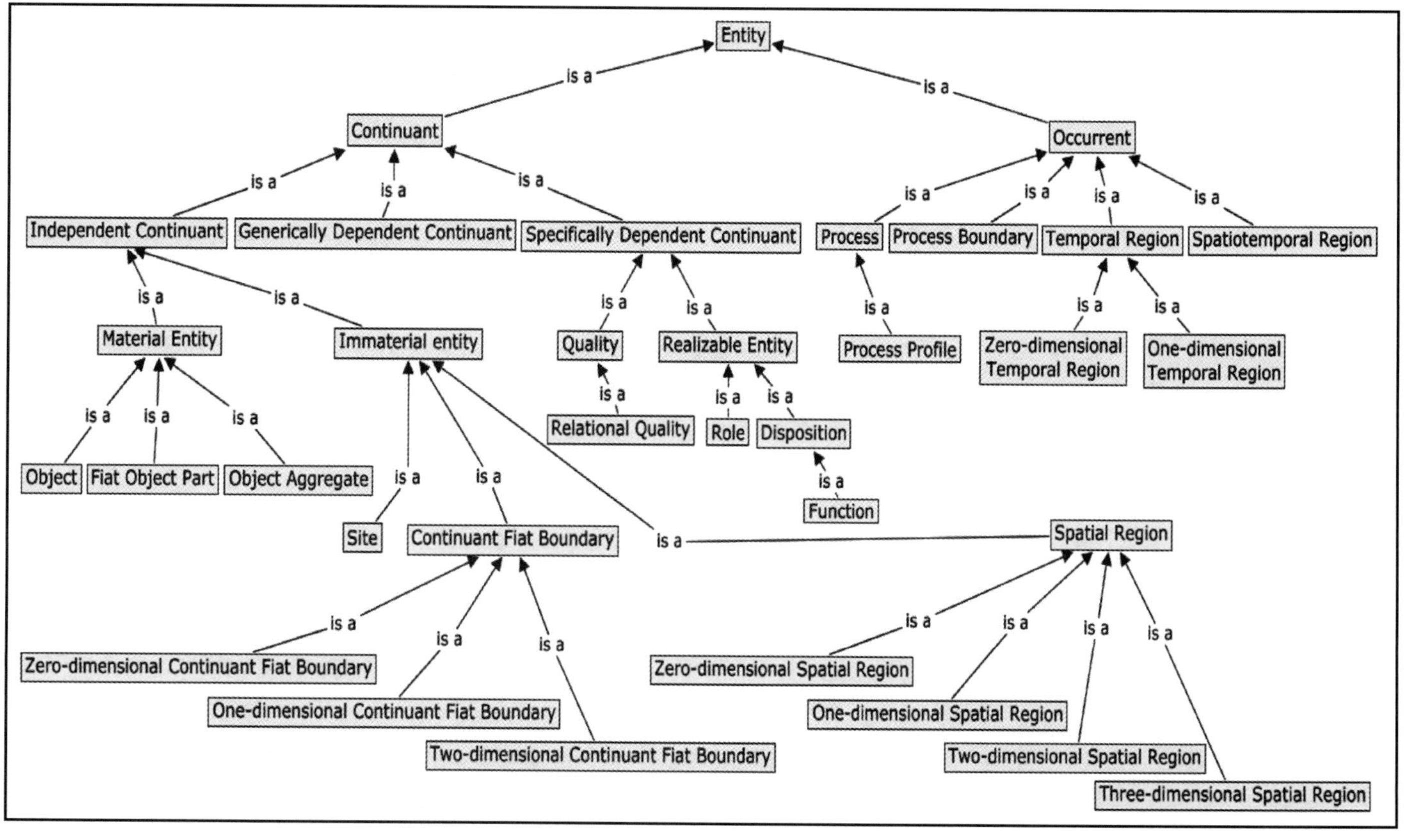

Figure 2. 34 Categories of the basic formal ontology. Source: permission of authors Robert Arp, Barry Smith, and Andrew Spear, *Building Ontologies with Basic Formal Ontology*, MIT Press, August 2015.

easily extended in a consistent and uniform way to ontologies having finer grained content as needed.[16]

The proposed framework extends from the basic formal ontology (BFO), which is a small, upper level ontology that is designed for use in supporting information retrieval, analysis, and integration in scientific and other domains. BFO is a genuine upper ontology designed to serve data integration in scientific and other domains.[17] It consists of only 34 classes, from which 13 will be employed in this framework—see Figure 2 for a full rendering of the BFO taxonomy. BFO adopts a view of reality as comprising: (1) continuants, entities that continue or persist through time, such as physical objects, qualities, and functions; and (2) occurrents, the events or happenings in which continuants participate. The continuant category is then further subdivided into those entities that exist independently, such as a person, and those entities that are ontologically dependent upon some other entity for their existence, such as some person's engineering or forgery capability. The subtypes of continuant and occurrent entities are represented in BFO as independent (material entity) and dependent (generally and specifically), continuant and occurrent.

The BFO provides the categories needed to decompose and analyze the complex OE. Analysis starts with decomposition, which is the process of separating some composite entity into constituent parts or elements. Decomposition exposes individual elements (nodes in a graph), and the relationships between them. The result is a semantic framework (i.e., ontology) that facilitates the organization of data about the OE. The semantic framework is populated with instance-level data that refers to the actual objects, realizable attributes, events, and relationships within the OE. Its purpose is to assist in the organization of data in such a way that it becomes useful for reasoning and decision making. The semantic framework will also help to identify gaps in the analyst's knowledge of the OE.

An independent continuant entity is a continuant entity that is the bearer of qualities (and realizable entities), and is a participant in occurrent processes. This category of entities includes physical objects that are the bearers of qualities, functions, roles, and dispositions—e.g., a person is an independent continuant entity that is the bearer of certain physical properties, and realizable attributes such as capabilities, functions, or roles. The independent continuant is represented as either a physical object (agent—person or organization) or object aggregate, such as a network.

Another type of independent continuant entity includes geographic features, which are further subdivided into natural and human-constructed geographic features. They are important for this framework because they are where clandestine networked cells operate. The data and information about geographic features will be addressed later in the chapter (see the section below on generically independent continuant—information artifacts). Independent continuant geographic features are represented as natural (hill, mountain, canyon, valley), or humanly-constructed (crops, industry, marketplace, city) geographic features.

The next major category in BFO is the dependent continuant entity. This category includes entities that are ontologically dependent upon independent continuant entities (i.e. their bearers) for their existence—i.e., they inhere in, or are borne by, other entities. Dependent continuant entities are further subdivided into generically dependent continuants (GDC) and specifically dependent continuants (SDC), and will be described later in the chapter.

Generically dependent continuant entities can migrate from one independent continuant entity to another—e.g., a PDF file can be transferred or copied to several hard drives simultaneously. The physical byte array may be different, but the generic content or meaning migrates with each instance of the PDF. For example, your jpeg image of Bill Clinton may be physically distinct from my exact copy, but the identical image migrates with each copy. We will return to GDC (information artifacts) later in the chapter.

SDC entity is defined as a continuant entity that depends upon precisely one independent continuant entity for its existence. The former is dependent on the latter in the sense that, if the latter ceases to exist, then the former will, as a matter of necessity, cease to exist also.[18] SDC entities exist as an attribute of an independent continuant, and cannot migrate to another independent continuant—i.e., an independent continuant is the bearer of some SDC. SDC entities include qualities and realizable entities such as capabilities, emotions, functions, and motivations. Qualities are exhibited, manifested, or realized in some independent continuant entity. They are represented in BFO as physical characteristics (color, height, mass, weight).

A realizable entity is defined as a SDC entity that has at least one independent continuant as its bearer, and whose instances can be realized (manifested, actualized, executed) in associated processes.[19] Examples of SDC entities include the roles of people in an organization or the dispositions of people that are realized by their actions and interactions. A disposition is defined as a realizable entity (a power, potential, or tendency) that exists because of certain features of the physical makeup of the independent continuant (e.g., a person) that is its bearer.[20] Examples include the capabilities, emotions, functions, and motivations that inhere in people and organizations. Examples of SDC realizable include: capability, emotional disposition, functions, and roles.

Emotional dispositions (action tendencies) and behavioral motives are properly treated in affective science, which is the study of emotions and of affective phenomena such as moods, affects, and bodily feelings. It combines the perspectives of many disciplines, such as neuroscience, psychology, and philosophy.[21] The BFO treats emotional dispositions (action tendencies) and behavioral motivations as BFO: realizable dispositions, which inhere in some person. These dispositions are realized in (BFO: occurrent) mental processes such as emotion occurrents and appraisals, which are addressed next.[22]

The next BFO category pertains to events, processes, happenings, or activities. An occurrent process is something that happens, or unfolds, through time and space, and always depends on some material entity. Recall that dependent continuant entities, such as roles, functions, or capabilities, are realized through occurrent processes. Naturally occurring processes include such things as geologic processes, weather events, biological processes, and planetary orbits. However, this chapter is concerned with another type of occurrent process, wherein agents (persons or organizations) are the participants. Actions are events that people perform with intentions and for reasons. One and the same action can be specified as intentional under some description and as purely physical under another description. But in order to be an action an event must have at least one description under which it is specified as intentional.[23]

A transnational terror-crime nexus can be represented as part of the complex operational environment. The ontological framework assists the analyst's understanding of the terror-crime nexus, and its place in the complex OE, by facilitating the creation of nodes and links (relationships) in a graph-like structure—a prerequisite for decomposition and analysis. This results in enhanced reasoning and information management. Figure 3 represents the basic ontology of a networked cell as the bearer of some realizable attributes, and is a participant in some occurrent processes (actions), which has location some geopolitical territory.

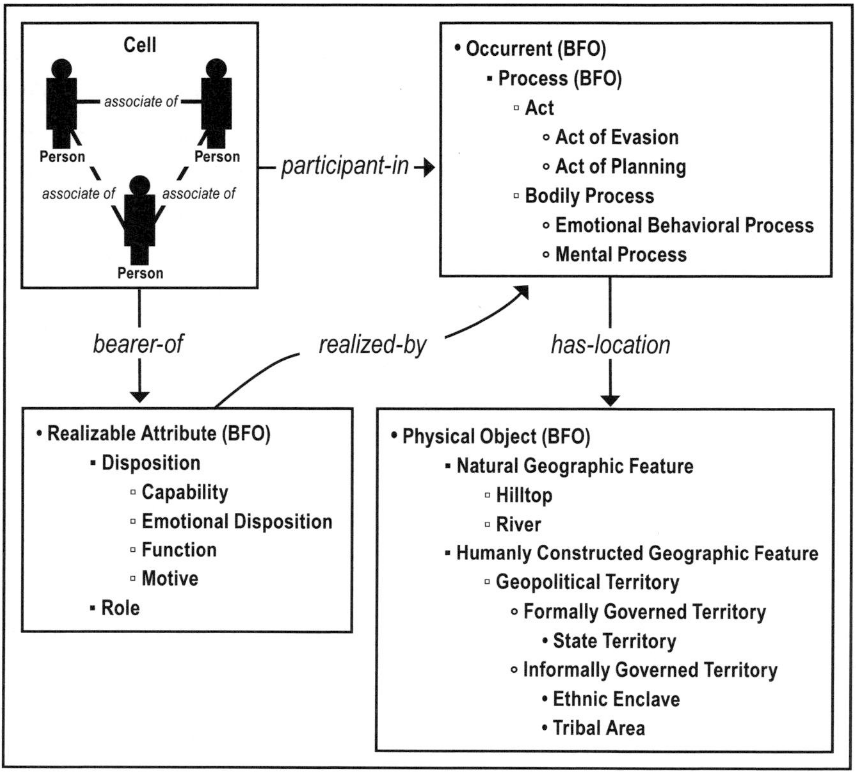

Figure 3. A basic ontology of cellular networks. Source: Author

The final BFO category to be considered in this chapter pertains to the data and information that is about the terror-crime nexus and the complex OE. GDCs, or information artifacts, can exist in (i.e., migrate between) more than one object at the same time—e.g., a certain pdf that exists in several different hard drives at the same time. The story of *War and Peace* migrates to every physical instantiation to include paper copies or digital versions in someone's reader. Information artifacts (e.g., descriptions, plans, grid coordinates, or images) are the subject of their own information artifact ontology, which is currently being developed.

Joint Intelligence Preparation of the Operational Environment

Data that is created, organized, or aligned with the above described ontological framework will be in a better format for JIPOE, which is the analytical process used by joint intelligence organizations to produce intelligence assessments, estimates, and other intelligence products in support of the joint force commander's decision-making process.[24] As previously described, JIPOE analysts adopt a systems perspective that is intended to provide an understanding of significant relationships within interrelated PMESII, and other systems relevant to a specific joint operation.

The systems perspective, that is representable with nodes and links in a graph, facilitates understanding of the continuous and complex interaction of friendly, adversary, and neutral systems.[25] Products developed during this step include network nodal and link analysis diagrams associated with adversary and neutral PMESII and other systems—see Figure 4 below. Some of the nodes will inevitably be linked to nodes in other PMESII systems, which may indicate a center of gravity—e.g., an influential political figure who controls access to land with valuable resources. Though there may be a wide variety of interactions between networks, dependency focuses on the scenario in which the nodes in one network require support from nodes in another network.[26]

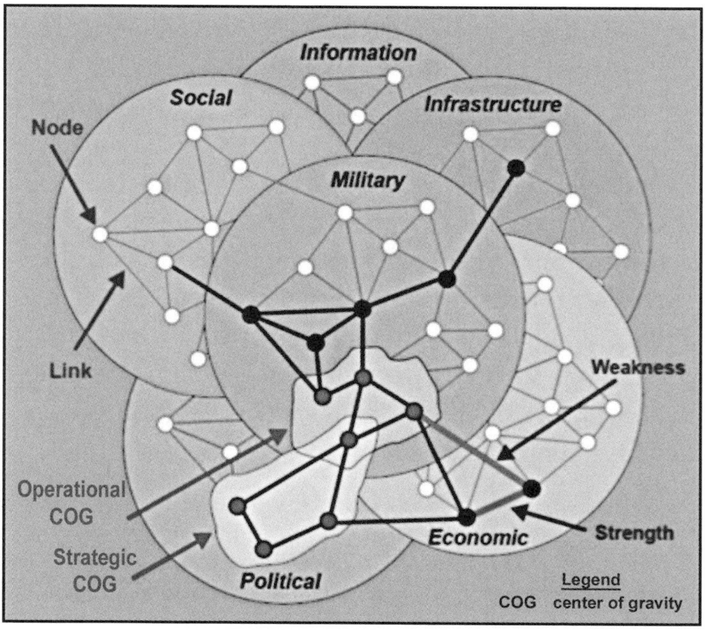

Figure 4. The Systems Perspective of the Operational Environment. Source: Joint Publication 2-01.3.

This data will enhance intelligence analysis at both the operational and strategic levels, as analysts study the nodes and links related to criminals, crime suspects, incidents, issues, and trends. By collecting and assessing this data, they can identify relationships or connections between different crimes in different places—e.g., suspect names, organization names, event locations, activity types, or even capabilities can show connections hiding within the available data. The analysis can then be used to provide insights that can drive or support CT operations and strategy, provide timely warning of threats, and influence government policy and decisions.

Conclusions: Applying the BFO Framework to Information Management Processes

The ontological framework described is intended to; (a) facilitate the CT analyst's ability to understand and describe the nexus between transnational terrorist organizations and criminal organizations, and (b) enhance the management of disparate data as it becomes available to the analyst, planner, or decision maker in CT operations.[27] This framework is applicable to situations where CT analysts have access to already existing structured data sets on terrorist activities and related events—this data will most likely be stovepiped in an IT system. It also applies to unstructured data from near-real-time tactical level reports, such as unit situation reports, or site exploitation team forensics reports, which contain detailed information from technical and forensic examination of documents, cell phones, computers, biometric information, weapons, bomb making, and other materials. Unstructured data consists of text strings, descriptive sentences, paragraphs, technical terms, and images. With some effort, this data can be 'tagged' with category terms taken directly from the ontological framework, making it interoperable with any other data that is similarly aligned—i.e., any data that is enhanced by the ontological framework becomes part of one large data set.

Endnotes

1. John Rollins and Liana Sun Wyler, *Terrorism and Transnational Crime: Foreign Policy Issues for Congress* (Congressional Research Service, 2013), 1.
2. Nils Gilman, Jesse Goldhammer, and Steven Weber, "Deviant Globalization," in *Convergence: Illicit Networks and National Security in the Age of Globalization* eds. Michael Miklaucic and Jacqueline Brewer, (Washington: National Defense University Press, 2013), 3.

3. National Security Council, *Strategy to Combat Transnational Organized Crime*, retrieved 09 July 2015 at: https://www.whitehouse.gov/administration/eop/nsc/transnational-crime/threat.
4. For more information on enhanced intelligence data see: http://ncorwiki.buffalo.edu/index.php/Distributed_Development_of_a_Shared_Semantic_Resource.
5. In a series of e-mail exchanges to the author on 18 March 2015, Ron Rudnicki described an overview of the Common Core Ontologies being developed for the Intelligence Information Warfare Directorate (I2WD).
6. U.S. Army, Army Training Publication 3-57.5, *Civil Information Management*. (Washington, D.C.: Government Printing Office, May 2013), 3-4; also see Appendix B, "Civil Analysis Example of the Interrelationship of Operational Variables and Civil Considerations;" B1-8.
7. Ibid., 6-7.
8. Robert Arp, Barry Smith, and Andrew Spear, *Building Ontologies with Basic Formal Ontology* (Cambridge, MA: MIT Press, August 2015).
9. U.S. Joint Chiefs of Staff, JP 3-07.2, *Antiterrorism* (Washington, D.C.: JCS, November 24, 2010), vii.
10. "Strategy to Combat Transnational Organized Crime: Definition" last modified search 12 May 2015, https://www.whitehouse.gov/administration/eop/nsc/transnational-crime/definition.
11. *Nexus*, Oxford English Dictionary, 2015.
12. John Rollins and Liana Sun Wyler, Terrorism and Transnational Crime: Foreign Policy Issues for Congress (Congressional Research Service, 2013), summary.
13. Derek Jones, *Understanding the Form, Function, and Logic of Clandestine Insurgent and Terrorist Networks: The First Step in Effective Counter-Network Operations* (Tampa: Joint Special Operations University Press, 2012), 73.
14. U.S. Joint Chiefs of Staff, Joint Publication 3-0, Joint Operations (Washington, D.C.: Joint Chiefs of Staff, August 11, 2011), xv-xvi.
15. U.S. Joint Chiefs of Staff, Joint Publication 2-01.3, *Joint Intelligence Preparation of the Operational Environment* (Washington, D.C.: Joint Chiefs of Staff, June 16, 2009), xii.
16. In a series of e-mail exchanges to the author on 18 March, 2015, Ron Rudnicki described an overview of the Common Core Ontologies being developed for the Intelligence Information Warfare Directorate (I2WD).
17. "Basic Formal Ontology" last accessed 22 June 2015, http://ifomis.uni-saarland.de/bfo/.
18. Arp, Smith, and Spear, *Building Ontologies with Basic Formal Ontology*, 95.
19. Ibid., 98.
20. Ibid., 101.

21. Richard Davidson, Klaus Scherer, H. Hill Goldsmith, Handbook of Affective Sciences, vol. Series in Affective Science, (New York: Oxford University Press, 2009).

22. Janna Hastings, Werner Ceusters, Barry Smith, Kevin Mulligan, "Dispositions and Processes in the Emotion Ontology" (paper presented at ICBO: International Conference on Biomedical Ontology, Buffalo, NY, 28-30 July 2011).

23. "Action Theory," The Internet Encyclopedia of Philosophy, accessed 22 June 2015 at: http://www.iep.utm.edu/davidson/.

24. U.S. Joint Chiefs of Staff, Joint Publication 2-01.3, *Joint Intelligence Preparation of the Operational Environment* (Washington, D.C.: U.S. Joint Chiefs of Staff, June 16, 2009), xi.

25. Ibid., xii.

26. Alessandro Vespignani, "Complex Networks: The Fragility of Interdependency," Nature (2010) 464 (7291): 984–985.

27. For more information on enhanced intelligence data see: http://ncorwiki.buffalo.edu/index.php/Distributed_Development_of_a_Shared_Semantic_Resource.

Chapter 11. Thoughts on Special Operations Forces Roles in Combating Transnational Organized Crime

Mr. Randy Paul Pearson

A Vexing Question

The question "can SOF have a role in countering transnational organized crime?" is a vexing one. On the one hand, the very inclusion of the word 'crime' in the question would lead one to logically conclude that this is obviously a law enforcement matter, not a military one. On the other hand, TOC is a rapidly morphing challenge that nobody can address alone. In fact, there is now a U.S. Presidential Strategy and Executive Order that expands national level efforts beyond the Department of Justice, to include the Departments of State and Treasury. Some of the enabling guidance also includes support from the DOD. SOF can and most likely will help to counter TOC drawing upon a rich history of capabilities not used in traditional law enforcement, yet falling well short of the traditional war-hammer used when SOF are committed to major combat operations.[1]

SOF are a unique military capability that can keep us to the left of war in a world growing in complexity and small-scale conflict. Because of their expertise, SOF are the perfect tool for such direct actions that keep the balance tilted toward peace without the high costs of large-scale war. While SOF can make valuable contributions as a surgical force application, their most enduring impact is through their adroit use of its well-honed concepts and analytical methods. These SOF capabilities can often be found

---

Randy Pearson is a national security professional with over 20 years of military, foreign affairs, policy, strategy, and intelligence experience. He currently advises U.S. Northern Command's counternarcotics program and is an Adjunct Lecturer for graduate studies in interagency relations and critical infrastructure protection at the University of Colorado at Colorado Springs School of Public Affairs. Mr. Pearson entered federal service as a Presidential Management Fellow at the U.S. Department of State.

in the best practices of the business community, operational law enforcement approaches, and in the routine modern processes of other government agencies. These specialties can be applied to existing problem sets to: help align guidance and resources; identify vulnerable nodes, protect those that are ours and exploit those that are not; and possibly develop a new way of orienting all elements of national power to countering the TOC threat. But criminal organizations can be resourceful too; indeed they have evolved into diverse transnational organizations far different from the structured organization of the Capone days.

Beyond Capone

One of the most storied celebrities in American organized crime history is that of Chicago gangster Alphonse Gabriel Capone. His connections and personal ambition fueled his rise to power, at a time of government intolerance for the mafia and for alcohol. The Italian government's crackdown on the mafia encouraged that population to take flight. The Capones landed in the United States, where Prohibition would eventually create the market in which Alphonse would thrive. The vexing question for law enforcement in his day was, "how could he be tied to his illicit activities?" On paper—as it were—Capone had no income or possessions. Similarly, he was far removed from any criminal activity, such as the Saint Valentine's Day Massacre, through proxies and alibis. But the discovery of an income ledger with Capone's name on it led to his prosecution and conviction for tax evasion.

One should note a few important items relating to the takedown of Al Capone that are relevant here. First, unlike most modern TOC networks, Capone's organization was regional and decidedly hierarchical. This lent itself to a template of easily understood identification and placement on an organizational chart for law enforcement to use. This in turn drove a successful decapitation strategy, although with short-term effects. While Capone was convicted and served time, eventually 'retiring' to Florida, "his organization continued for fifty years after the end of Prohibition," by diversifying into other illicit markets and practices.[2] TCOs are much more resilient and can readily match supply with demand, diversifying sources and supply chains with a snap. They are complex with multidimensional veins, breeding an air of immortality. The modern approach to countering this threat is more often than not the application of a dated singular decapitation strategy vice that of a multi-pronged interagency nodal targeting campaign. This results in the continuation of the criminal organization beyond the incarceration of its bosses or the interdiction of its soldiers. In effect, the end state is a disruption, without much further consideration for detection, denial, or dismantlement. Thus, the terminology used in planning guidance is vital to set the correct orientation to arrive at the desired destination.

Second, it was forensic accounting by the IRS that discovered Capone's finances. This work, provided by another government agency, enabled Capone's arrest by law enforcement, judicial prosecution, and eventual conviction. One could then argue that the key to cracking the Capone case lay within the toolbox of a non-law enforcement organization. The IRS had a resident talent that it was willing to contribute to a larger effort—setting a precedent for interagency cooperation.

Finally, much like his contemporaries of today, Capone was well insulated from his association with his illicit trade. Until the discovery of the fatal ledger, his name was never associated with any income or possession. This remains a best practice of the illicit trade. The difference between then and now is that globalization and technological innovations have enabled illicit networks to further insulate their membership with regeneration capabilities and anonymity. This allows for a complex exploitation of relationships and infrastructures that transcend normative boundaries or singular analytical talents to combat illicit activities. To counter the illicit networks and the activities of TOC, the U.S. Government has developed policy and strategy to be implemented with a whole-of-government, indeed cooperative international approach.

Language Alignment

Words have meaning—one reason most documents have a glossary or terms of reference section. If everybody doesn't understand the terminology used, it is more probable than not that the result of their work will produce unintended consequences.

The end state of the 2011 SCTOC is to reduce TOC "from a national security threat to a manageable public safety problem in the United States and in strategic regions around the world."[3] On that same page is the key policy objective to "defeat transnational criminal networks." That same bullet includes the terms "depriving" and "preventing," both of which are synonymous with 'deny' rather than 'defeat.' Circling back to the desired end state, the term 'defeat' does not necessarily correspond to the word or meaning of 'reduce.' One could imply that this displays the complexity that is TOC. One could also imply that the strategy conflates the means to an unintended end.

In the FY15 National Defense Authorization Act, Congress made the attempt to make counternarcotic (CN) activities more flexible when assisting in the countering of TOC. This is important for the purposes of this discussion because CTF is a complimentary activity to the DOD CN program with a carefully constructed range of tasks and metrics reporting requirements. In constructing the guidance for CTF, the DOD nested its guidance from the NSS down the chain through something currently known as the DOD Directive for Counter Threat Finance Policy, which was last revised in 2012. This directive states that DOD policy is "to deny, disrupt, or defeat and degrade adversaries' ability to use global licit and illicit financial networks."[4] Note the inclusion of the term "defeat" in the policy.

An interesting occurrence may be the latest updates to two different joint publications in which SOF play a role. Joint publications are revised every few years to keep the documents current with their communities and ever evolving guidance. They go through an extensive review process. The first publication of note is JP 3-24, *Counterinsurgency*, from November 2013. This update defines CTF as "operations [that] may be conducted to disrupt and deny finances or shut down networks."[5] The second publication of note is the 2014 JP 3-26, *Counterterrorism*. This update defines CTF as "an interagency effort to detect, counter, contain, disrupt, deter, or dismantle the transnational financing of state and non-state adversaries threatening U.S. national security."[6]

The point of this exercise in definitions is that 'defeat' may be recognized as a bridge too far for what may be considered the 'shaping of the environment' or 'phase zero' activities. This could be considered the wrong phase in which to execute the mission of defeating an adversary, which is usually reserved for 'seize the initiative' (phase three) or 'dominate' (phase four) activities.[7] Furthermore, joint publications are an entrée into something called the Universal Joint Task List (UJTL). This list "is a menu of tasks in a common language, which serves as the foundation for joint operations planning across the range of military and interagency operations."[8] It essentially tells DOD components what they are supposed to do at the strategic national, strategic theater, operational, and tactical levels, and how to measure those efforts. Nowhere in this March 2015 task list is the term 'defeat' assigned to CTF (or CN) tasks.

While the main focus of SOF training and equipping are core activities, "SOF can also perform collateral activities such as counterdrug operations."[9] In the conduct of counterdrug activities (which includes CTF), the objectives of SOF are not to 'defeat' any adversary. Rather, SOF operate in a joint, multinational, or multiagency environment, providing training to mission partners according the UJTL.[10] SOF can assist in the updating and continued development of TOC and CN guidance in at least two main ways. First, SOF, through USSOCOM, can advise on guidance reviews by advancing the criteria of the UJTL back up through the chain to ensure that nested documents have aligned language. Second, SOF, through the conduct of both their core and peacetime collateral activities can help to solidify the concept that it is in support of other agencies or mission partners. These two approaches will help to establish not only the alignment of guidance to reference from, but to lower the barriers to interagency cooperation that can prevent objective accomplishment. Both are key in the alignment of objectives and arriving at the end state desired, vice the outcome unintended.

Priority Analysis

SOF are masters of target acquisition, and the supporting evidence to this end is substantial. Part of this discipline includes the ability to build the criteria for target selection and prioritization with an emotional detachment from the equation. In a fair amount of cases, with the right tools, the easy or common sense target is proven to be not so easy or common. And these tools are more diverse in application than they first appear. The origins of the SOF analytical tool called CARVER—criticality, accessibility, recuperability, vulnerability, effect, and recognizability—lays mostly within oral military history. However, as a tool it is well documented in doctrine and it is now used in a variety of other fields outside that of the military for its logic process and defensive applications.[11]

CARVER is a methodology used to rank critical vulnerabilities resulting in target prioritization. Its purpose is to identify systems and their tributaries that must be attacked in order to degrade an adversary's capacity. The CARVER matrix is split into a two-axis table with the CARVER criteria in columns across the top, and targeted systems listed down the left hand side. A system or subsystem is assigned a number (e.g., 3) according to its importance or vulnerability, which is drawn from a numerical value range (e.g., 1-5) assigned to each of the six criteria of CARVER. The totals of those numbers indicate the rank of targets with the largest numbers becoming the priorities.[12]

Nonmilitary organizations usually credit SOF with originating and developing CARVER before they explain their purpose and adaptation of it. These organizations include government entities such as the Departments of Energy and Homeland Security, as well as private sector organizations such as policy and cyber security firms. One predominant example is the Food and Drug Administration (FDA), which has wholly embraced the CARVER method and is likely to be among the top five Internet search engine results when looking up the term. The FDA flips CARVER over for the purpose of food security, adapting the method to identify vulnerabilities that require defensive measures or hardening.[13] In the FDA adaptation of this method a seventh criteria—"shock"—is added to evaluate "the combined health, economic, and psychological impacts of an attack."[14] This addition could become useful in interagency cooperation and further analysis of systems for its shock value against a target.

CARVER can have great impact with multiple applications. As an offensive targeting tool, it can be applied against adversary systems. As a defensive tool it can identify vulnerabilities to nodes in a supply chain, which assists with the prioritization of resource investment and decision making for mitigation. CARVER could also be applied in the logic chain of prioritizing universal joint tasks (e.g., coordinate, direct, dismantle, deter, detect, and or conduct with regard to CTF). However, if other government agencies were to adapt CARVER analysis in the whole-of-government approach to countering TOC, it could prove to be the key to identifying the most effective course of action against a target, and thus the appropriate element of national power to lead or execute that effort. In short, CARVER could assist the United States Government with prioritizing its investments and applying its resources, which saves the country money. And when stacking a $1 million CN budget up against that of multibillion narcotics trafficking enterprise, every cent counts.

Nobody Can do it Alone

TOC is so extensive that it requires context to even begin to explain its reaches. The 2010 UN TOC threat assessment reported:

> globalization has outpaced the growth of mechanisms for global governance, and this deficiency has produced just the sort of regulation vacuum in which transnational organized crime can thrive. People and goods can move more cheaply than ever before ... flows are too intense to easily distinguish the licit from the illicit ... And the rapid pace of change itself provides opportunities for organized crime.[15]

This presents a permissive global environment in which upwards of $3 trillion is illicitly laundered, $1 trillion worth of narcotics is smuggled, and human trafficking nets an estimated $21 billion.[16] With a global GDP estimated to be around $74 trillion, just the illicit proceeds listed above represent almost 6 percent of the world's GDP.

Context can also be applied against relative successes in combating organized crime. As with the above Capone example, the 2014 arrest of cartel boss, Joaquin "El Chapo" Guzman has not brought an end to the Sinaloa cartel. In fact, the cartel functions much the same way as before Guzman's capture, operating in at least 17 Mexican states and some 50 countries, spanning across the Western Hemisphere,[17] and quite probably across the globe. So, despite this long sought and celebrated arrest, the Sinaloa cartel has not been reduced to a 'public safety problem'—the desired end state of the 2011 TOC strategy. That doesn't mean it can't get there. Just as flipping the CARVER method over from an offensive to defensive application, aspects of globalization can also be flipped to leverage technologies and collaborative security partnerships in countering TOC. Examples include USSOCOM's Global Counterterrorism Network, the Global ECCO program, and the Congressionally funded Combating Terrorism Fellowship Program, which sustain networks of CT practitioners who can work collaboratively to counter TOC.[18]

The president's cover letter to the TOC Strategy states:

> this strategy is organized around a single, unifying principle: to build, balance, and integrate the tools of American power to combat

> transnational organized crime and related threats to our national security—and to urge our partners to do the same.[19]

The ability of SOF to build, whether it be concepts, processes, and relationships, is one of their greatest strengths. An essential SOF trait to building is that of adaptability—something government is largely not well known for. However, as retired General Stanley McChrystal illustrated in a recent *Business Insider* discussion about Uber, adaptability can be found and cultivated in many places, including the private sector. In essence, Uber has leveraged technology with a "digital backbone" to build a network of trust for a common purpose that has resulted in a "shared consciousness."[20] The lesson this concept may have for government agencies countering TOC is much the same. If the common purpose of reducing TOC to a public safety problem is the end state for interagency mission partners, SOF could have the role of partnering, or building the necessary networks of trust-based relationships, beyond the military and interagency arenas, and into the civil society. The opportunity to do just that could manifest itself in the collaborative work and technological leaps that are taking place right now, and could bring about a revolution in the basic constructs of how governments address TOC threats.

Leveraging the Cloud

When Hani Hanjour, Ziad Jarrah, and Mohammed Atta were the subjects of separate traffic stops prior to the 9/11 terrorist attacks, a dynamic communications or data query system to alert local law enforcement to FBI, Central Intelligence Agency (CIA) or even neighboring police information could have played the single preventative role to the terrorist attacks that changed the world.[21] A similar system might have helped FBI internal communications and maybe even piece together the swatches of information regarding concerns about Middle Eastern men taking flight lessons in Phoenix, Arizona.[22] The fact is there are lots of systems out there that could be leveraged not only for law enforcement purposes, but that of national security as well.

In January 1967, an information sharing system known as the National Crime Information Center (NCIC) went online for the first time. It contained 95,000 records from all 50 states in a centralized location. As of 2014, NCIC contains some 12 million records and set a record for processing over 14 million queries in a single day. It allows law enforcement partners to conduct off-line searches, instant criminal background checks, biometric queries, and other information sharing services.[23] Another data center, the National Security Agency's Bumblehive, in Bluffdale, Utah, was opened in 2014 and is rumored to have the ability to expand its storage capability to the incomprehensible yottabyte.[24] It is designed to store, match, and share data via a supercomputer that can process 100,000 trillion calculations each second.[25] And yet another system is the $600 million computing cloud that was developed by Amazon for the CIA and went online in 2014. It is designed to coordinate all 17 intelligence agencies, "allowing agencies to share information and services much more easily and avoid the kind of intelligence gaps that preceded the 9/11 terrorist attacks."[26]

Arguments against the use of supercomputing by governments state that such an elaborate capability places too much power in the hands of government. Taking several leaps further, these arguments can cast anchor into industrial era thinking at the very least, and nurse a universe of paranoid delusion at the very most. But this misses the mark, entrenching status quo, perverting information, and most of all allowing TOC to flourish. To move beyond Capone, CTOC efforts must be captured in the context of collaborative security, not a monolithic government worthy of no more than suspicion.

Cloud computing is designed for sharing. It makes it easy to 'plug and play' at minimal costs. Through the concept of "multi-tenancy architecture"[27] infrastructure, services, and applications are shared among all partners. This architecture is also scalable to the size or needs of an organization as needed (e.g., National Security Agency vs. the sheriff's office, or steady-state vs. earthquake response), and it can be customized for each partner's own purposes (e.g., CTOC vs. CT). Cloud computing is widely used in the private sector where customers place such trust in the concept that they have placed some of their most personal information on it.

The CIA computing cloud is particularly relevant here. If its fidelity and applicability allow for the secure processing and communicating of TOC-related information, it could potentially prove to be the basis for linking different levels of classified materials to push a releasable product or message. Similarly, all of these centers could be linked to share a common releasable (e.g., 'tear-line') product platform, or allow for a query system that could result in a similar product, or identify a positive hit in need of a further classified consultation.

This is where SOF could have a role exercising their talent of adaptability to help bring this idea or something like it all together. USSOCOM, through its Interagency Partnership Program, provides Special Operations Support Teams to mission partner organizations to synergize and accomplish mutually beneficial tasks. While NCIC, the Bumblehive, and the CIA's cloud all have the intent to share information, that information is still only shared with a select set of partners, with much of the information being pulled (vice pushed) from the data center. Just as with General McChrystal's Uber example, a Special Operations Support Team could help build the 'digital backbone' to the world's greatest threat information sharing architecture and bring about an age of unprecedented interagency collaboration. This in turn could bring about a 'shared consciousness' through supercomputing that enables government functions, investment prioritization, and application of resources to reduce TOC to a public safety problem.

Conclusion

There are lots of reasons to justify not doing something. As with Capone after Prohibition, it can take a long time to effect change and Americans are not fond of long timelines. TOC networks present a complexity that is not fully understood and is difficult to explain. They also have resources in exponential reserves beyond that of what government funding is applied against them. Government agencies also have barriers to cooperating with one another. Not only are there legal, procedural, investigative, and classification reasons, but the perception of a too powerful omniscient government all form to become a consensus for remaining in the status quo.

However, nobody can tackle TOC alone. While much of it remains a law enforcement matter, its growth through convergence and the effects of globalization make it a modern day Leviathan that is rapidly outpacing the resources that governments apply against it. SOF can have a role in CTOC by bringing their capabilities and talents to the table. This will not cause TOC to be militarized. To the contrary, because SOF remain in support of interagency mission partners, their application against TOC can stay well to the left of a costly full-scale war.

One of the five major recommendations of the 9/11 Commission Report was to unify "the many participants in the counterterrorism effort and their knowledge in a networked-based information sharing system that transcends traditional government boundaries."[28]

SOF's role could help interagency communications, analysis, and information sharing. SOF can help to align language across government guidance and definitions. SOF can help apply CARVER analysis to stated government actions and objectives to ensure that the appropriate government element takes the appropriate action to arrive at the desired end state. Finally, SOF can help build a whole-of-government information sharing enterprise that can be shared across all levels of government and law enforcement, possibly developing a new way of orienting all elements of national power to countering the TOC threat.

Endnotes

1. David S. Maxwell, "Thoughts on the Future of Special Operations," *Small Wars Journal*, 31 October 2013.
2. Frank G. Shanty, Organized Crime: From Trafficking to Terrorism, ABC-CLIO, Santa Barbara, California, 2008, 46.
3. Strategy to Combat Transnational Organized Crime: Addressing Converging Threat to National Security, The President of the United States, July 2011, 1.
4. DOD Counter Threat Finance (CTF) Policy, Department of Defense Directive 5205.14, 16 November 2012.
5. Joint Publication 3-24, *Counterinsurgency*, 22 November 2013, xvi.
6. Joint Publication 3-26, *Counterterrorism*, 24 October 2014, xi.
7. Joint Publication 3-0, *Joint Operations*, 11 August 2011, V-5 through V-9.
8. Universal Joint Task List (UJTL), Joint Electronic Library, 17 June 2015, as found at: http://www.dtic.mil/doctrine/training/ujtl_tasks.htm.
9. Joint Publication 3-05, *Special Operations*, 15 July 2014, II-2.
10. Joint Publication 3-07.4, *Counterdrug Operations*, 14 August 2013, IV-6. The Universal Joint Task List "is a menu of tasks in a common language, which serves as the foundation for joint operations planning across the range of military and interagency operations." It essentially tells DOD components what they are supposed to do at the strategic national, strategic theater, operational, and tactical levels, and how to measure those efforts.
11. Joint Publication 3-05.1, *Joint Tactics, Techniques, and Procedures for Joint Special Operations Task Force Operations*, 19 December 2001, V-17. "The SOF mission planning process relies on the ... (CARVER) matrix to determine the feasibility of and methodology for applying SOF to the target."
12. Army Doctrine Reference Publication 3-05, *Special Operations*, Chapter 4, 31 August 2012, 4-6 through 4-8.
13. Hardening is generally accepted as the act of applying security to a component of a system or infrastructure to improve its resilience.

14. Food and Drug Administration, Carver plus Shock Primer, September 2009, accessed at: http://www.fda.gov/Food/FoodDefense/FoodDefensePrograms/ucm376791.htm. FDA provides an overview, guides, and downloadable software for its adaptation of CARVER that can be found at: http://www.fda.gov/Food/FoodDefense/ToolsEducationalMaterials/ucm295900.htm.

15. The Globalization of Crime, A Transnational Organized Crime Threat Assessment, United Nations Office of Drugs and Crime, 2010, 29.

16. Transnational Organized Crime Foldout, National Intelligence Council Publications, Director of National Intelligence, 2011.

17. Sinaloa Cartel, In Sight Crime, n.d., as accessed on 27 June 2015 at: http://www.insightcrime.org/mexico-organized-crime-news/sinaloa-cartel-profile.

18. Global ECCO, accessed 1 July 2015 at: https://globalecco.org.

19. Ibid., TOC Strategy.

20. General Stanley McChrystal on how Uber operates just like a Special Forces unit, Business Insider, 15 May 2015, as found at: http://www.businessinsider.com/general-mcchrystal-manage-company-us-army-uber-2015-5.

21. Susan Candiotti, "Another hijacker was stopped for traffic violation," CNN.com, 9 January 2002, accessed at: http://edition.cnn.com/2002/US/01/09/inv.hijacker.traffic.stops/.

22. Scott Olster, "FBI's Phoenix memo unmasked," *Fortune*, 12 October 2011, accessed at: http://fortune.com/2011/10/12/fbis-phoenix-memo-unmasked/.

23. Criminal Justice Information Services, 2014 Annual Report, US Department of Justice, accessed at: https://www.fbi.gov/about-us/cjis/annual-report-2014/cjis-annual-report-2014.

24. A yottabyte is the eighth power of 1000. The multiple less is a zettabyte, which is the seventh power of 1000. The estimated example of what a zettabyte might look like is the equivalent of all of the grains of sand on all of the Earth's beaches.

25. Domestic Surveillance Directorate, Utah Data Center, National Security Agency, 2014, accessed at: https://nsa.gov1.info/utah-data-center/.

26. Frank Konkel, "The Details About the CIA's Deal with Amazon," *The Atlantic*, 17 July 2014, accessed at: http://www.theatlantic.com/technology/archive/2014/07/the-details-about-the-cias-deal-with-amazon/374632/.

27. Multi-tenancy is an architecture in which a single instance of a software application serves multiple users.

28. The 9/11 Commission Report, Final Report of the National Commission on Terrorist Attacks Upon the United States, W. W. Norton & Company, Inc., 400.

Conclusion: SOF Roles and Future Challenges

Chapter 12. World Order or Disorder: The SOF Contribution

Mr. Michael Miklaucic

As the era of the 'big footprint' winds down and the U.S. relies increasingly on allies and partner nations, it is troubling that so many states are fragile or mired in seemingly never-ending conflict. Nearly 25 years after the Cold War, no fewer than 65 of the 193 member states of the UN are considered as 'high warning,' 'alert,' or 'high alert' by the 2015 Fragile States Index, while only 53 are rated 'very sustainable' through 'stable.' What this tells us is that the set of states capable of effective security partnership is quite small.

What's the Problem?

State fragility or failure are endemic and invite a range of illicit actors, including international terrorists, globally networked insurgents, and TCOs. Their presence and operations keep states weak and incapable of effective partnership. Illicit organizations and their networks both engender and feed

---

Michael Miklaucic is the Director of Research, Information, and Publications at the Center for Complex Operations (CCO) at National Defense University and the Editor of PRISM, the CCO journal. He served in various positions at the U.S. Agency for International Development and the Department of State, where he served as the Deputy for War Crimes Issues. In that position he was responsible for U.S. relations with the International Criminal Tribunal for Rwanda, the Special Court for Sierra Leone, war crimes issues and negotiations in East Timor and Cambodia, and the early implementation of the Sudan Peace Act. He is adjunct professor of U.S. foreign policy at American University and conflict and development at George Mason University.

corruption that erodes state legitimacy and the confidence of the governed in the state as the legitimate guardian of the public interest. These networks penetrate the state, leading to state capture, and even criminal sovereignty. A growing number of weak, failing, captured, and criminal states are creating gaping holes in the global rule-based system of states that we depend on for our security and prosperity.

The last 10 years have seen unprecedented growth in interactivity amongst a wide range of illicit networks, as well as the emergence of hybrid organizations that use methods characteristic of both terror and crime. In a convergence of interests, terrorist organizations collaborate with cartels and trafficking organizations collude with insurgents. International terrorist organizations, such as al-Qaeda, deprived of state funding for their operations, have adapted by engaging energetically in transnational crime to raise funds. Prominent criminal organizations like Los Zetas have adopted the brutal symbolic violence of terrorists—the propaganda of the deed—to secure their 'turf.' And networked insurgents, such as the FARC, or the Liberation Tigers of Tamil Eelam, mutated from insurgency to both crime and terror.

The unimpeded trajectory of these trends—state capture, convergence, and hybridization—poses substantial risks to the national security interests of the U.S. and threaten international security. Illicit networked organizations are challenging the fundamental principles of sovereignty that undergird the international system. Fragile and failing states are at the same time prey to such organizations, which feed on them like parasites, and act as homes for them, incapable of supporting effective security partnerships. The Westphalian, rule-based system of sovereign polities itself is at risk of fraying over the long run, as fewer capable states survive to meet these challenges, and populations around the world lose faith in the Westphalian paradigm.

The Westphalian system is relatively new, and though well-tested and resilient over several centuries, its long-term survivability cannot be assumed. Alternative futures are imaginable. This chapter describes the phenomena I refer to as 'convergence,' the interactivity and hybridization of diverse illicit networks, and the conventional remedies employed by the international community to counter these trends. The chapter then examines dystopian visions of a world in which these trajectories go indefinitely unimpeded. It concludes by discussing emerging and disturbing new networks and patterns of interactivity and possible countermeasures to be explored.

Clash and Convergence

The conventional wisdom informs us that TCOs and international terrorist organizations are unlikely candidates for partnership. Criminals, driven by the venal pursuit of wealth in defiance of law, morality, or ideology, have little interest in the struggles, violence, or risks taken by international terrorists. The very last thing they desire is to be pursued by Navy SEALs or the CIA. International terrorists and insurgents, on the other hand, are driven by ideological or nationalistic motives, and repelled by the vulgar materialism and rampant greed of criminals. They have no desire to get on the radar of the DEA or other national or international law enforcement agencies. This logic is understandable, indeed rational, and may have prevailed in previous times, but the present chapter will argue that the evidence of extensive interconnectivity—if not explicit partnership—between TCOs, international terrorists, and globally networked insurgents is compelling. If the separation of these diverse types of organizations and networks was ever as completely hermetic as the conventional wisdom suggests, that separation is a quaint anachronism from a bygone era.

Yet there are still some who remain skeptical of these linkages despite a growing literature exposing such connectivity. Regardless of the nature of their motivations and relationships, evidence suggests these interactions have reached unprecedented levels. Recent research undertaken at the Combating Terrorism Center at West Point reveals that, "criminals and terrorists are largely subsumed (98 percent) in a single network as opposed to operating in numerous smaller networks."[1] The six degrees of separation that once may have divided people is a relic of the past—international terrorists, insurgents, and criminals are merely a telephone call from each other.

Though the evidence of their connectivity is now overwhelming, we remain largely in the dark regarding the nature of their agreements or arrangements. We lack the telephone intercepts or written documents describing these connections. Some relationships are better understood than others. For example, it is known that in 1998 Ayman al-Zawahiri's Egyptian Islamic Jihad organization merged with Osama bin Laden's al-Qaeda, because the two organizations respectively wished for the new relationship to be known. Other relationships, such as between the FARC and al-Qaeda in the Islamic Maghreb (AQIM) are opaque because neither organization has an interest in revealing the relationship. According to a DEA spokesman, roughly half of the Department of State's 59 officially designated foreign terrorist organizations have been linked to the global drug trade.[2]

Are they merging or working in partnerships? Are their partnerships simply 'marriages of convenience,' or 'one-time deals?' There is some risk in mirror-imaging—i.e., expecting these illicit networks and organizations to mimic the kinds of relationships we find among legal, formal organizations. We often hear that illicit organizations operate like ordinary businesses motivated by similar sets of incentives. I disagree with that characterization. There is a fundamental qualitative difference between companies motivated by the quest for profit through legal means, and organizations that cheat, extort, even murder to achieve their goals, whether those goals are monetary or ideological. Though terrorists, insurgents, and criminals may operate under codes of conduct, 'agreements among thieves,' or other informal rules of engagement, we should not expect to find codified relationships or agreements. They operate clandestinely under constant threat of exposure, capture, or death. Illicit networks and organizations operate outside of our paradigm.

Nothing New Under the Sun

Some argue that terrorism, insurgency, and organized crime have existed throughout history, and that their modern iterations represent nothing unprecedented. This seems naive—modern enablers such as information and communication technology, transportation advances, and fabulous volumes of money are game-changers. They permit illicit actors to avail themselves of lethal technology, military-grade weaponry, real-time information, and professional services of the highest quality, including legal, accounting, technological, and security services. Cartels and gangs, as well as terrorists, and some insurgents can now outman, outspend, and outgun the formal governments of the countries where they reside. Illicit actors can communicate across the globe in real-time, using widely available and inexpensive technology. The November 2008 Mumbai terrorist attackers used satellite phones, Internet communications, and global positioning systems, under the direction of Pakistan-based handlers.[3] Furthermore international travel has never been easier or cheaper than it is today. Would-be terrorists, traffickers, launderers, even assassins can fly from continent to continent nearly undetected in the sea of traveling humanity. This was not the case in the past.

But How Big of a Problem is This?

Part of the challenge in understanding illicit organizations and networks is their purposeful opacity. Operating by definition and intention outside the vision of regulators or researchers, their activities and revenues are hidden. So how do we determine the magnitude of their operations, or the harm they cause? How do we know the value of their transactions? We extrapolate from extremely inexact evidence, such as seizures, arrests, convictions and the associated testimony of witnesses, often themselves members of such organizations and motivated to dissemble.

When estimating the dollar value of global illicit markets, many commentators rely on the now nearly 20-year-old International Monetary Fund (IMF) consensus range; the figures $1 to 3 trillion, or 2 to 5 percent of global product have been circulating since 1998 when Michel Camdessus, then managing director of the IMF, gave that estimate for the amount of money laundered globally each year. Given what we know about global trafficking in drugs, persons, weapons, counterfeits, and other contraband, it seems unlikely that the dollar value of illicit trade has decreased over the past 20 years. Even at a mere 2 to 5 percent of global product, Camdessus described the magnitude of the problem nearly 20 years ago as "almost beyond imagination ..."[4]

Less difficult but more visceral to calculate is the cost in human lives of global terrorism. The year 2014 saw an increase of 35 percent in the number of terrorist attacks globally, with total fatalities rising to nearly 33,000 by some counts,[5] not to mention non-fatal injuries, the destruction of families and communities, and the economic costs. These cannot be monetized, but few would deny that the opportunity cost of the Global War on Terrorism (GWOT) has been huge. The Iraqi Freedom component of GWOT alone has been estimated to have cost as much as $3.5 trillion, or nearly 6 percent of global product.

Simply for illustrative purposes, if we add these two sets of costs using these estimates—the global illicit market plus the costs associated with the Iraqi Freedom portion of GWOT—the sum is $6.5 trillion, or nearly 11 percent of global product. Admittedly these estimates are notional and there is no claim to scientific accuracy, but they seem credible—and staggering. Just consider the opportunity cost if one-tenth of human activity is dedicated to transnational crime and terrorism. Then add to this the cost of networked insurgencies to such countries as Afghanistan, Colombia, Sri Lanka, and South Sudan.

Why is this Our Problem?

Converging illicit networks pose a direct challenge to sovereignty, rendering states unable to effectively govern their territory or borders, let alone exercise a monopoly of the legitimate use of force, or provide other vital public services. A cursory examination of a few key states shows the toll illicit networks take on our national security interests. The cost is imposed at three levels: the inability of states to govern their populations and territories creates seedbeds for international terrorism, networked insurgency, and transnational crime, causing immense human suffering; state fragility and instability frequently have a regional spillover effect, sometimes penetrating key U.S. allies and partners; and growing feral regions serve as launch pads for attacks on U.S. national security interests, as well as potentially direct attacks on the homeland, such as occurred on 9/11.

Some of our most important national security partners and potential partners are states in critical condition largely due to the imprecations of illicit networks. Though Mexico's death rate has subsided considerably over the past two years, the wars between the narcotics cartels and the state authorities, and between the cartels themselves, are thought to have caused over 60,000 deaths between 2006 and 2012, or an average of nearly 10,000 per year. With our most populous immediate neighbor experiencing casualties like that, it is no wonder Mexico has been an exporter of insecurity. Mexican cartels today work hand-in-hand with the criminal gangs of Central America's Northern Triangle—El Salvador, Honduras, and Guatemala—which as a result is experiencing some of the highest homicide rates in the world. El Salvador's official forensic unit estimates the homicide rate in 2014 at nearly 70 per 100,000. Despite their collaborative intentions these countries are under such duress that their security partnership contributions cannot yet inspire confidence. Indeed, last year 50,000 unaccompanied children from Central America made their way through Mexico seeking entry to the United States, presenting a significant foreign policy challenge.

Nigeria is a key potential security partner for the U.S. in Africa. The most populous African nation with the largest economy, and a major oil exporter, Nigeria could and should play a stabilizing role throughout the continent.

Nigerian forces were critical in staunching the civil wars that hemorrhaged West Africa in the 1990s and 2000s. Yet today Nigeria is hobbled by the burgeoning Boko Haram insurgency in the north, and the resurgent gang insurgency in the Niger Delta region. Moreover, the Boko Haram scourge has bled into neighboring Niger and Cameroon, and its shrouded connections with al-Shabaab in Somalia and AQIM in Mali threaten a continent-wide insurgency.

At various times both Iraq and Afghanistan have been spoken of as U.S. national and international security partners. Indeed as noted above, the United States invested trillions of dollars to bolster the capacity of these two partners, yet today it is hard to imagine either state as an effective security partner. Afghanistan today struggles to survive the combined attacks of al-Qaeda, the Taliban, and Haqqani networks, and now the Islamic State. Although the government of Afghanistan remains a partner in the sense that it welcomes U.S. engagement, its effectiveness as a security partner remains questionable. The future of Iraq is equally unclear as the Iraqi military fights the Islamic State, but depends on Kurdish and Shi'a non-state militias. As of this writing, it is impossible to say that either Afghanistan or Iraq has been stabilized such that it will not provide a base for terrorist groups planning attacks against the United States in the near future.

But Things are Getting Better, Right? Not!

The death spiral in Mexico appears to have subsided—for the moment. Colombia has recovered from its near-death experience at the end of the 20th century, and is today even an exporter of security. But these are isolated cases—proof of concept that effective remedial action can reverse the assault on sovereignty by converging illicit networks. In other regions things are getting worse, not better.

Above, I mentioned connections between Boko Haram in Nigeria, and both al-Shabaab in Somalia and AQIM in the Sahel region. Though the nature or extent of these connections is not transparent, what is clear is that when these groups desire to communicate, collaborate, coordinate, and collude, they are able to do so. Joint training, learning, and sharing of experience are certainly likely, if not yet joint operations. Moreover, the known connection between al-Shabaab and al-Qaeda, and allegedly to the Islamic State, provides an Asia-Africa terrorism conduit, while the AQIM-FARC relationship extends the conduit from Africa to the Western Hemisphere. Converging illicit networks pose a direct challenge to sovereignty, rendering states unable to effectively govern their territory or borders, let alone exercise a monopoly of the legitimate use of force, or provide other vital public services.

The Northern Triangle of Central America, composed of Guatemala, Honduras, and El Salvador, has become a highway for traffickers, primarily of drugs from the producing areas of South America to the consuming areas of North America—but also of weapons, humans, and other illicitly traded commodities. Criminal gangs, originally formed in the prisons of California, have proliferated and metastasized throughout the sub-region, bringing with them their unique brand of tattooed violence, resulting in the world's highest homicide rates. In El Salvador, a 2012 truce agreement between the incarcerated chiefs of MS-13 and Calle 18, resulted in a substantial reduction in the homicide rate, begging the question, "Who exercises sovereign power in El Salvador?" How and with whom can we forge an effective security partnership?

Equally if not more disturbing than the growing power of non-state illicit groups is the collusion between such groups and rogue elements of sovereign states, such as Pakistan's Inter-Services Intelligence directorate, long known to support the Taliban and Haqqani networks. Iran's Quds Force, a special forces unit of the Islamic Revolutionary Guards Corps, has been both directly engaged in terrorist acts around the world, and supportive of other terrorist organizations. More ominously in 2011, the Quds Force attempted to smuggle an assassin into the United States for the purpose of killing the Saudi Arabian ambassador in a plot referred to as the "Iran assassination plot."[6] The most disturbing aspect of this case was the attempt by the Quds Force to collaborate with the Los Zetas cartel organization in this effort. That this plan was intercepted by the vigilant DEA is extremely fortunate—at this particular moment in time, with the extreme tension between Iran and Saudi Arabia, and between Sunni and Shi'a throughout the Islamic world, the consequences of the intended assassination are difficult to imagine. One need only consider the consequences of a diplomatic assassination that took place in Sarajevo a century ago to put this into perspective.[7] This effort by the Quds Force to collaborate with Los Zetas, now fully documented in U.S. case law, demonstrates beyond a reasonable doubt, the potential collusion of terror organizations with criminal organizations.

There are in addition certain states apparently willing to tolerate and even directly engage in criminal and terror activity, utilizing the toolkit of international statecraft in the effort. Venezuela, for example, has utilized diplomatic pouch privileges, passport issuance, and a variety of other diplomatic tools to support criminal and terror activity.[8] North Korea has long been known as a hub of illicit activity allegedly including smuggling, counterfeit trade, production of controlled substances, illegal weapons trafficking, and money laundering. Pyongyang's infamous Room 39 is thought to generate between $500 million and $1 billion per year from such illicit activities.[9]

Could Things Get Any Worse?

As comforting as it would be to believe things cannot get any worse, such faith would be naive. Today the Islamic State's assault on Syria and Iraq is being vigorously resisted by a coalition that includes many American partner countries, including Saudi Arabia, the United Arab Emirates, Jordan, Qatar, and Bahrain, among others. Imagine a world in which the U.S. had no capable partners. American troops would be required wherever and whenever a national security interest was threatened overseas. No partners in Central America would mean that should state failure in the region result in a threat to the control of the Panama Canal, U.S. troops would be required to restore or maintain safe transit through the canal. No capable partners in Africa would mean U.S. boots on the ground to combat al-Shabaab, Boko Haram, and AQIM.

Already the world has closed in to an extent on the domain of freedom. Many countries in which it was safe for a Westerner to travel only a generation ago, are now off limits to prudent travelers. Among the 67 casualties of the Westgate Shopping Mall terrorist attack in Kenya in 2013, no fewer than 19 were foreigners of different nationalities. Likewise, the 2008 terrorist attacks in Mumbai, India, resulted in the deaths of 28 foreign nationals from 10 countries, out of the total of 164 dead. Though statistically such events should not necessarily deter anyone from visiting Kenya or India, taking appropriate precautions, they indicate a world in which it has become less safe for Americans to travel.

There is also the vital question of global trade and commercial activity. Total trade (exports and imports) accounted for 30 percent of U.S. GDP in 2013, supporting 11.3 million jobs.[10] For the most part, we trade with our partners. When a partner succumbs to the corruption and social morbidity of illicit networks, its economic capacity is diminished, rendering it a less valuable trading partner. Imagine the growth potential of a Central America free of gang violence and cartel trafficking.

Is There Any Hope?

If the 20th century was consumed by the global struggle between incompatible ideologies, fascism, communism, and democratic capitalism, the 21st century will be consumed by the epic struggle to create and sustain viable and effective states. Viable, effective states are the only form of collective governance that has a proven ability to contain and reverse a trajectory of growing entropy driven in part by illicit networks. States have successfully fought off powerful illicit adversaries in all regions of the globe, from Colombia in South America to the Philippines in Asia. Some authors have argued that the state itself is a significant contributor to growing global entropy, and that is likely true.[11] Yet effective, viable states have enabled great prosperity and security, and alternatives to state-based governance are few.

State building regrettably has been discredited over the past two decades.[12] Both the cost and difficulty inherent in a realistic approach to state-building have soured policymakers and budget-setters to the proposition of trying to stand up states. Indeed the epic state-building failures in Iraq and Afghanistan, where budgets were unprecedentedly large, have proven to many the futility of the effort and the concept. It is true that coalition efforts to build effective, viable states in Iraq and Afghanistan failed, and before that were the archetypal failures in Somalia and Haiti. These failures were not inevitable. There are numerous examples of successful state-building efforts, including South Korea, The Republic of China (Taiwan), Singapore, and Colombia (a successful case of state-rebuilding), among others. Many of these began their trajectory toward democracy with periods, even prolonged periods, of autocratic rule—a fact we must seriously consider.

We do know from considerable experience that state building is an arduous, labor-intensive, and time consuming task. There is extensive literature on the subject, and widely diverging views on how it should be done, but virtual unanimity regarding the intensiveness of the process.[13] What does state-building consist of? Though far from a science—still more alchemy than chemistry at this stage—there are a few principles that draw wide agreement.

Above all, the state must establish a secure and stable environment for public and private life. Famously prescribed by Max Weber as the singular defining attribute of a state, "a state is a human community that (successfully) claims the monopoly of the legitimate use of physical force within a given territory."[14] Importantly, Weber specified the legitimacy of the use of force. No state has or can enjoy a complete monopoly of the use of force—nor would we necessarily want it to. However for the use of force to be legitimate it must be sanctioned by the state. Historically the state use of force has been conceived as a responsibility to protect citizens from external aggressors, though in many cases the state itself has been an aggressor. This unpleasant reality has recently been addressed by a growing acknowledgment that the state's responsibility for security extends beyond its own survival to its population—hence the emerging concept of 'human security.'

Establishing a monopoly of the legitimate use of force in a territory is no mean feat, and cannot be accomplished by brutal methods without sacrificing the legitimacy that is essential to effective governance. There are numerous U.S. Government programs that provide assistance, training, equipping, mentoring, and other supports to partner governments, both military and civilian agencies, for the purpose of BPC. No amount of training and equipping, however, can substitute for the social contract between government and governed necessary to establish and sustain legitimacy. This must be achieved by our partners. In this our role can only be to help them identify methods, techniques, best practices (to the extent we know them), and lessons to enable their success. Controlling the use of force within sovereign territory, either directly or through delegation, is an essential function of a viable state.

The application of force in society must be bound by the rule of law, another critical responsibility of the state. The state must establish the rule of law and mechanisms for articulation, adjudication, and redress of grievances. Doing so provides methods for the resolution of social and other disputes within society, provides predictability necessary for commerce, and ensures the security of citizens. The rule of law is not just a question of constitutions or statutes, though they form the legal framework in any country. It also requires that citizens have access to the law and the institutions of justice, and that they are not excluded from legal recourse by cost, language, distance, or identity. Under a genuine rule of law, the state itself is also subject to the law and cannot operate outside the law or the legal system. This includes prohibiting and punishing corruption especially within government agencies both civilian and military.

In order to execute its required functions effectively and sustainably the state must invest in and upgrade human capital. It must create a reliable and competent civil service to administer official state functions and manage state assets. It must also provide a social and security environment conducive to education and public health. A professional civil service, including administrators, diplomats, and a range of support personnel necessary to operate complex systems is required to assume the full spectrum of governmental responsibilities.

The state must develop systems, human resources, and institutional mechanisms for raising revenue, securing state financing, and managing state assets. No state can operate effectively without a stable and predictable revenue stream sufficient to meet the costs of its obligations. There is controversy over what a state's obligations are and vast variation among states; but whatever obligations compose the social contract between government and governed must be within the financial means of the state. Taxation and regulation of commercial and financial activity is a responsibility only appropriate for the state. A banking system capable of interaction in the global financial network is beyond the capabilities of the private sector.

The degree to which the state must or should be involved in commercial activity is debatable, and indeed subject of wide historical debate that has generated large and powerful intellectual and political schools of thought on the subject. There is, though, a degree of consensus on the state's responsibility with respect to creating an environment in which citizens' economic needs are met. The debate is over the balance of responsibility for meeting those needs within a conducive environment. According to the prevailing contemporary wisdom, the state should enable individual economic and commercial innovation, bearing only a modest responsibility beyond that through taxation and limited redistributive programs.

Critical to state sustainability is an inclusive national narrative promoting citizenship. While not the exclusive responsibility of the state, typically only the state has access to the nationwide communication systems required to disseminate strategic messaging about national issues. The drafting and adoption of a national constitution can contribute to an inclusive national narrative, as can elections. These can be divisive, but the craft of statesmanship is being able to manage and utilize such formal processes in support of the national interest. The willingness of citizens to pay taxes to the state is contingent upon an inclusive national narrative and a social contract that citizens accept. An effective state penetrates most aspects of public life, including education, public health, commerce, and dispute resolution among many others. It can use those platforms to further the forging of an inclusive national narrative. This requires high standards of leadership, without which no state will succeed in any case.

The state must also create accessible mechanisms for interaction between civil society and the state. Robust civil society encourages associative behavior conducive to social capital and enables citizens to pursue their interests equitably amidst the competing interests within any state. The state cannot form or create civil society, but it can communicate and interact transparently and responsively with civil society organizations. Furthermore it can provide and secure the political space needed for their operations.

While this is merely a notional short-list of state responsibilities and functions (of which there are many more), and described in a summary manner, my intent is to emphasize the centrality of the state to sustaining the rule-based world order. A world without order is a frightening prospect that recalls Hobbes' characterization of the natural state of mankind before government is established as "every man against every man," and a life that is "solitary, poor, nasty, brutish, and short."

SOF to the Rescue

What does all this have to do with us? SOF have unique capabilities and strengths that can help meet the sovereignty challenges facing our partner states. In many of these states the military is the institution most respected by citizens, and considered most capable. In some cases, armed forces as a subset of the population are better educated than nonmilitary cohorts. They typically consist of individuals who have made a commitment to serve the country. According to retired Admiral Dennis Blair, "The great majority of officers first put on the uniform to protect their country and its citizens ..."[15] The armed forces importantly have the critical mass necessary to accomplish ambitious goals that extend throughout the state.

What is often lacking in our partner states is a collegial and congenial relationship between citizens and their own armed forces. Armed forces are frequently cantoned away from the general population, and form an isolated community within their own society. SOF are trained to work with civilians, and can tap that strength, working with their counterparts to build skills required to work effectively with civilians. Linda Robinson, of the RAND Corporation, writes that one strength, "is developing and working alongside indigenous forces to combat terrorists, insurgents, and transnational criminal networks through an orchestrated set of defense, information, and civil affairs programs."[16]

While supporting partner armed forces and law enforcement agencies in neutralizing key criminal, insurgent, and terrorist leaders, SOF can also help these armed forces build bonds with society at large. Robinson states:

> special operations forces may engage in nonlethal activities such as dispute resolution at the village level, the collecting or disseminating of information, or civil affairs projects such as medical or veterinary aid and building schools or wells. Persuasion and influence are part of many of these operations, and the long-term effect is to build relationships and partnerships that endure.[17]

What she doesn't write, but is consistent with the principle of BPC, is the value of transferring the skills associated with civilian-military relationships to our partner forces. Helping our partners build relationships and their own civilian-military partnerships that endure is one of the ways in which SOF can help meet the challenge of strengthening states.

A Lot to Lose

The Westphalian system of autonomous, sovereign states, interacting according to a set of universally understood norms and rules have always been at best an aspiration—never fully accomplished even in Europe, let alone the world beyond. Yet the concept of a rule-based system of sovereign states has contributed to a world in which successfully consolidated and integrated states have flourished. In the 367 years since the Peace of Westphalia established this rule-based system based on sovereign equality, the world has experienced an unprecedented surge across a range of quality of life indicators; life expectancy has surged from below 40 to over 70 years; per capita GDP increased from around $600 to over $10,000 per year; literacy has increased from less than 10 percent to over 80 percent of the global population.

It may be that other organizing principles for political activity are superior to the Westphalian state, but such alternatives are not evident. Throughout previous millennia political life was governed by such governing structures as tribes, clans, fiefdoms, kingdoms, empires, cities-states, and religions. These organizing principles did not prove to be effective or competitive in the modern world, and in most regions gave way to state structures. In the 20th century Vladimir Lenin tried to establish a political order based on economic class, with catastrophic effect.

For lack of a better alternative, the rule-based system of states in which the U.S., among many others, has flourished in recent centuries, is well worth saving. Its current condition should be a matter of profound concern among any interested in the 'long-term.' Converging illicit networks threaten that system, and must be countered if the system is to survive the current generation intact. The first step must be to recognize and acknowledge the nature and magnitude of the threat.

Endnotes

1. Scott Helfstein with John Solomon, "Risky Business: The Global Threat Network and the Politics of Contraband," Combating Terrorism Center at West Point, 16 May 2014, accessed at: https://www.ctc.usma.edu/v2/wpcontent/uploads/2014/05/RiskyBusiness_final.pdf.
2. Remi L. Roy, "Dissecting the Complicated Relationship Between Drug Operations and Terrorism," The Fix, 8 October 2014, accessed at: http://www.thefix.com/content/dissecting-confounding-nexus-drugs-and-terror.
3. Jeremy Kahn, "Mumbai Terrorists Relied on New Technology for Attacks," *New York Times*, 8 December 2008, accessed at: http://www.nytimes.com/2008/12/09/world/asia/09mumbai.html?_r=0.
4. "Money Laundering: the Importance of International Countermeasures," Address by Michel Camdessus, Managing Director of the International Monetary Fund, Plenary Meeting of the Financial Action Task Force on Money Laundering, Paris, 10 February 1998, accessed at: https://www.imf.org/external/np/speeches/1998/021098.htm.
5. "Country Reports on Terrorism 2014," U.S. Department of State, June 2015, accessed at: http://www.state.gov/documents/organization/239631.pdf.
6. Department of Justice, Office of Public Affairs, Press Release, 30 May 2013, accessed at: http://www.justice.gov/opa/pr/manssor-arbabsiar-sentenced-new-york-city-federal-court-25-years-prison-conspiring-iranian.
7. On 28 June 1914, Gavrilo Princip, a Bosnian Serb, assassinated Archduke Franz Ferdinand of Austria, precipitating a chain of events that lead to World War I.

8. Douglas Farah, "Terrorist-Criminal Pipelines and Criminalized States: Emerging Alliances," *PRISM*, V.2, No.3, 2011.
9. Kelly Olsen, "North Korea's secret: Room 39," Associated Press, 11 June 2009, accessed at: http://www.sltrib.com/ci_12566697.
10. Stephanie Han and Natalie Soroka, "U.S. Trade Overview, 2013," International Trade Administration, U.S. Department of Commerce, October 2014, accessed at: http://www.trade.gov/mas/ian/build/groups/public/@tg_ian/documents/webcontent/tg_ian_002065.pdf.
11. Randall Schweller, *Maxwell's Demon and the Golden Apple* (Baltimore: Johns Hopkins University Press, April 2014); see also, https://jhupbooks.press.jhu.edu/content/maxwells-demon-and-golden-apple.
12. Dominic Tierney, "The Backlash Against Nation-Building," *PRISM* V.5,N.3, Summer 2015, accessed at: http://cco.ndu.edu/Publications/PRISM/PRISMVolume5,no3.aspx.
13. See Clare Lockhart and Ashraf Ghani, *Fixing Failed States* (New York: Oxford University Press, 2009); Francis Fukuyama, *State-Building* (Ithaca, NY: Cornell University Press, 2004); and James Dobbins, *The Beginner's Guide to Nation Building* (Santa Monica, CA: RAND, 2007), for a sampling of this literature.
14. Max Weber, "Politics as a Vocation," lecture to Free Students' Union, Munich 28 January 1919, accessed at: http://anthropos-lab.net/wp/wp-content/uploads/2011/12/Weber-Politics-as-a-Vocation.pdf.
15. Dennis Blair, "Military Support for Democracy," *PRISM* V.3, N.3, Summer 2012, accessed at: http://cco.ndu.edu/Portals/96/Documents/prism/prism_3-3/prism3-16_blair.pdf.
16. Linda Robinson, "The Future of U.S. Special Operations Forces," Council on Foreign Relations, Special Report No. 66, April 2013.
17. Ibid.

Acronym List

AQIM	al-Qaeda in the Islamic Maghreb
ASCOPE	areas, structures, capabilities, organizations, people
BFO	basic formal ontology
BPC	building partner capacity
CAF	Canadian Armed Forces
CANSOF	Canadian Special Operations Forces
CANSOFCOM	Canadian Special Operations Forces Command
CARVER	critically, accessibility, recuperability, vulnerability, effect, and recognizability
CDS	Chief of Defence Staff (Canada)
CFDS	Canada First Defence Strategy
CIA	Central Intelligence Agency
CICTE	Inter-American Committee Against Terrorism
CN	counternarcotic
COE	contemporary operating environment
CSIS	Canadian Security Intelligence Service
CT	counterterrorism
CTF	counter threat finance
CTOC	counter transnational organized crime
DEA	Drug Enforcement Administration
DHS	Department of Homeland Security
DOD	Department of Defense
FARC	Revolutionary Armed Forces of Columbia

FATA	Federally Administered Tribal Areas
FATF	Financial Action Task Force
FBI	Federal Bureau of Investigation
FDA	Food and Drug Administration
FID	foreign internal defense
FINTRAC	Financial Transaction Reports Analysis Center of Canada
FY	fiscal year
GDC	generically dependent continuants
GDP	gross domestic product
GWOT	Global War on Terrorism
HSBC	Hong Kong and Shanghai Banking Corporation
IMF	International Monetary Fund
IRS	Internal Revenue Service
ISIL	Islamic State of Iraq and the Levant
ISIS	Islamic State of Iraq and Syria
JIPOE	Joint Intelligence Preparation of the Operational Environment
JP	Joint Publication
JSOU	Joint Special Operations University
JTF	Joint Task Force
MND	Minister of National Defence (Canada)
NATO	North Atlantic Treaty Organization
NCIC	National Crime Information Center
NSC	National Security Council
NSP	National Security Program

NSS	National Security Strategy
NYPD	New York Police Department
OAS	Organization of American States
OE	operational environment
ONDCP	Office of National Drug Control Policy
PMESII	political, military, economic, social, information, infrastructure
RCMP	Royal Canadian Mounted Police
SCTOC	Strategy to Combat Transnational Organized Crime
SDC	specifically dependent continuants
SNA	social network analysis
SOCNORTH	Special Operations Command North
SOF	Special Operations Forces
SOLA	Strategic Orientation Look Ahead
SOUTHCOM	United States Southern Command
TCO	transnational criminal organizations
TEU	twenty-foot equivalent units
TFI	Terrorism and Financial Intelligence (U.S. Department of Treasury)
TFOS	Terrorist Financing Operations Section
TOC	transnational organized crime
UJTL	Universal Joint Task List
UN	United Nations
USSOCOM	United States Special Operations Command
USSOF	United States Special Operations Forces